More Praise for

See What I'm Saying

"To cover the range of our extraordinary perceptual skills, [Rosenblum] provides fascinating, concrete examples for each ability. . . . [An] appealing and compelling look at new findings about the powers of our less-conscious brain, the realm of the senses."

—Whitney Scott, *Booklist*

"[Rosenblum's] writing is always engaging."

—Matt Chorley, *Popular Science*

"Psychologist Rosenblum (Univ. of California, Riverside) describes in language accessible to lay readers a quirky collection of sensory wonders, which he then explains scientifically and also describes how to duplicate easily. . . . Fans of Steven Pinker's *How the Mind Works* will find a cousin in this science book for nonscientists." —*Library Journal*

"In accessible language, Rosenblum guides us to novel first-hand experiences, then explains the science behind them. Particularly fun is when he puts himself under the microscope. . . . *See What I'm Saying* demonstrates that the five senses do not travel along separate channels, but interact to a degree few scientists would have believed only a decade ago. After reading Rosenblum's captivating book, you will be surprised at how much your senses are capable of."

—Richard E. Cytowic, professor of neurology, George Washington University, in *New Scientist*

See
What
I'm
Saying

The Extraordinary Powers
of Our Five Senses

Lawrence D.
Rosenblum

W. W. Norton & Company

New York London

For information about permission to reproduce selections from this book,
write to Permissions, W. W. Norton & Company, Inc.,
500 Fifth Avenue, New York, NY 10110

For information about special discounts for bulk purchases, please contact
W. W. Norton Special Sales at specialsales@wwnorton.com or 800-233-4830

Manufacturing by Courier Westford
Book design by Lovedog Studio
Production manager: Julia Druskin

Library of Congress Cataloging-in-Publication Data

Rosenblum, Lawrence D.
    See what I'm saying : the extraordinary powers of our five senses /
Lawrence D. Rosenblum.—1st ed.
        p. cm.
    Includes bibliographical references and index.
    ISBN 978-0-393-06760-6 (hardcover)
1. Senses and sensation. I. Title.
    BF233.R67 2010
    152.1—dc22
                                                        2009047975

ISBN 978-0-393-33937-6 pbk.

W. W. Norton & Company, Inc.
500 Fifth Avenue, New York, N.Y. 10110
www.wwnorton.com

W. W. Norton & Company Ltd.
Castle House, 75/76 Wells Street, London W1T 3QT

1 2 3 4 5 6 7 8 9 0

For Bita, Mateen, and Kian

with gratitude and love

Contents

Part IV. TOUCHING

Part V. SEEING

Part VI. MULTISENSORY PERCEPTION

Preface

WHAT ARE YOU PERCEIVING RIGHT NOW? YOUR EYES SEE THE words printed on this page. Your hands feel the textures of the paper and cover of this book. Your ears hear nearby and more distant sounds. Your nose might detect food or the laundry detergent on your shirt or the soap on your skin. Your tongue might taste the salt, sugar, or other remnants left over from your last meal. As you introspect further, you'll notice less prominent things: the visual background adjacent to this book, the feeling of the clothes on your skin, the sound of your own breathing.

But at this very moment, you are accomplishing a number of perceptual feats that are more difficult to experience—and more difficult to believe. You are hearing things that don't make noise. You are feeling things that don't touch your skin. You are smelling things that have no discernible odor. You are seeing things that have no form. And, you do these things all the time. In fact, these exotic perceptual skills are critical to your connection to the world and to your survival.

These more exotic skills are all based on the physical information available to your senses along with the physiology that allows your senses to take in this information. These perceptual skills are not clairvoyance or mindreading, or supernatural in any way. They are real, replicable, and scientifically knowable.

New research in perceptual psychology and brain science is revealing that our senses pick up information about the world that we thought was only available to other species. But, as science is learning, we can use sound like bats, smells like dogs, and touch like bugs, and we do so constantly. As our more-conscious brains are busy with the trivialities of our day, our less-conscious brains are engaged in much more interesting endeavors. Our less-conscious brains are absorbing a profusion of sights, sounds, and smells using processes that seem superhuman. While psychologists have long known that our sensory systems can take in information without our awareness, new research is showing that entire perceptual skills are ocurring this way. These *implicit perceptual abilities* are allowing our less-conscious brains to have all the fun. A goal of this book is to bring some of that fun to our more-conscious selves.

To help you understand your sensory superpowers, this book introduces you to individuals with highly developed perceptual skills. You will meet a blind man who uses his inborn sonar system to lead mountain bike expeditions and others who use the sounds of a beeping baseball to hit and field with astonishing accuracy. You'll learn how an expert lip-reader is able to perceive speech as well as, and from a greater distance than, a hearing person, and how a deaf-blind individual can perceive speech by *touching* the face of a talker. You'll observe how wine connoisseurs taste the vintage of an obscure French wine, and how a fragrance designer can expertly match an aroma to the visible décor and clientele of a hotel. You'll hear how a champion a cappella group maintains perfect pitch, and how a blind surveillance expert can recognize phone numbers by listening to the complex pitches produced by telephone touch tones. You'll learn how an expert fisherman can feel the type, sex, and age of a fish on his line, and how a professional food taster can discern how long it will be before a cookie's ingredients will begin to turn. Importantly, these examples will not be used to argue that only some individuals are perceptually gifted but rather to highlight the potential of *everyone's* perceptual powers. We *all* have an onboard sonar system and a type of absolute pitch; and we all can perceive speech from seeing and even touching faces. What's more, we engage many of these skills all day long. What largely distinguishes the expert perceiver from the rest of us is the same thing that gets us from here to Carnegie Hall: practice. By understanding the expertise of others, you'll get ideas on how to improve your own implicit perceptual abilities.

The new discoveries of perceptual superpowers are certainly exciting in their own right, but these discoveries are also scientifically important in revealing emerging principles of perceptual psychology and brain science. For example, recent evidence that sensory compensation occurs in not just sensory-impaired individuals, but in all of us, supports a growing conception of *neural plasticity* that is far-reaching. It turns out that we all have brain regions and cells that can change their function depending on our experiences. Furthermore, the growing evidence for our ability to use multiple senses for what until very recently were considered single-sense functions (perceiving speech from *seeing* faces; perceiving a person's attractiveness from *smells*) supports the emerging notion that the brain is designed around multisensory input. In some ways, the brain doesn't much care which sense organ provides information. This fact is even true of the supposed "visual" and "auditory" brain centers that, we now know, incorporate multisensory input. Finally as you'll discover, many of our implicit perceptual skills involve sensitivity and reactivity to *human* actions. Extraordinary sensitivity to what others are *doing* is consistent with evidence for brain cells that have the dual role of recognizing and *producing* an action (throwing, smiling). As we silently listen to speech, cells that are involved in initiating our speech *movements* are activated. These *mirror systems* may help explain our tendency to automatically mimic aspects of each other's behaviors such as facial expressions and speech inflections. In sum, as you read about your implicit perceptual skills, you'll also learn about the important new principles of perceptual and brain science that explain these skills.

So why haven't you heard about these implicit perceptual skills before? And why can this book be written now? The answers lie with some recent advances in perceptual science.

It is a particularly good time to be a brain. Contemporary technologies for seeing the brain at work (functional magnetic resonance imaging, evoked potentials, magnetoencephalography, transcranial magnetic stimulation) have allowed for a number of critical discoveries about perception. For example, the technologies have revealed that in important ways, your brain knows more about the outside world than you do. New evidence suggests that your brain is constantly reacting to stimuli for which you have little conscious awareness. As an example, you'll learn in Chapter 4 that your brain is activated by odors long after you have habituated to and no longer notice them. It turns out that much of what your nose does, it does unconsciously. The new research is also showing that your brain doesn't simply *react* to unconscious input, it uses this input in meaningful ways that bear on your judgments and behavior.

New imaging technologies also show that your brain is much more ecumenical when it comes to your individual senses than once thought. As mentioned, areas of your brain once assumed to be dedicated to a single sense actually help out with multiple senses. This indicates a more sophisticated early treatment of perceptual input than has been assumed. The imaging techniques have also allowed for discoveries of brain areas reactive to both initiating and perceiving the same actions (mirror systems). These findings are consistent with what is now known of our intricate yet often unconscious skills to mimic one another (Chapter 9). Observations of these perception-action brain systems are also consistent with research showing that you are best at perceiving aspects of your environment that have direct behavioral relevance: You perceive the world in terms of what you can do with it.

But, one of the most important discoveries in neuroscience over the last 20 years is that the brain can change its structure and organization as it's influenced by experience. The degree of this neuroplasticity comes as an exciting surprise to a science that long assumed that once mature, relatively little changes in the brain's structure. Brain regions once thought dedicated to particular perceptual functions have the potential to rededicate themselves to different functions both within and across the senses. And as you'll learn in Chapter 6, this brain restructuring and reassignment can occur from a wide range of experiences running from simple finger taps, to the cross-modal plasticity occurring with sensory loss of either the permanent (blindness) or quite temporary (90 minutes of blindfolding) type. Observing how the perceptual brain changes with experience has informed our understanding of perceptual expertise and how it is established. All of these provocative observations about the brain would not be possible without the powerful imaging techniques that have become widely available over the last 10 years.

Not only is it a good time to be a brain, it's also a good time to be a perceiving, acting animal. A significant shift in how perceptual phenomena are understood and evaluated has occurred. This new approach prioritizes an understanding of the external information available to our senses over the internal mental processes typically thought to reconstruct the external world. This *ecological* approach to perception attempts to explain how animals perceive and act in their natural environments. And it assumes that once considered in their natural context, our senses can access robust information available in the physical patterns of light, sound, etc. Your senses swim in seas of rich information that help establish what you (implicitly and explicitly) know about the world and how you can behave toward it. In a way, you're surrounded by what you know.

A natural by-product of the approach is the discovery of perceptual sensitivities that are not typically considered. The ecological approach argues that once humans are observed in a context of sufficiently rich, natural sensory information—and as soon as they are asked to perform tasks that have behavioral relevance—previously unobserved perceptual skills emerge. In this sense, the approach predicts exotic perceptual abilities. And ecological perception research has revealed a myriad of exotic perceptual skills. These skills include your ability to *hear* the shape of both sound-making and sound-reflecting objects (Chapters 1 and 2); to feel the size and shape of an object simply by shaking a small piece of it back and forth (Chapter 7); and to visually recognize people simply based on the way they move (Chapters 8 and 9). These are just a few of the examples showing human sensitivity to highly rich perceptual information not typically considered.

Thus, this book has emerged out of the fortunate convergence of new observations from brain imaging techniques and an important shift in how perceptual questions are being asked. The complementarity of brain and perceptual methods is useful for addressing most questions in psychology. After all, to understand how the brain works, it's critical to understand what the brain is for.

■ ■ ■ ■ ■

See What I'm Saying will introduce your hidden perceptual skills through stories of expert perceivers, discussions of new research findings, and simple demonstrations. The goals of the book are threefold: (1) to reveal that you have extraordinary perceptual skills of which you are largely unaware; (2) to show that by becoming aware of these skills and how they work, you can actually enhance how to use them; and (3) to convey that these skills illustrate some of the most important recent developments in perceptual science.

Acknowledgments

So MANY PEOPLE HELPED WITH THIS BOOK, IN SO MANY WAYS. IN the beginning, my mother and late father, Lenore and Milton Rosenblum, inspired the general topic of this book (and my career) through their work with sensory-impaired individuals. Of course, their constant love and support didn't hurt either. The dry wit of my brother, Rob Rosenblum, helped in developing the voice of this book. Michael Turvey, Carol Fowler, and Alvin Liberman provided superb academic guidance and helped refine my thinking on many of these topics.

As the idea for this book took shape, Ken Rosenfeld and Peter Del Greco were some of the first and most encouraging supporters. Sonja Lyubomirsky provided constant inspiration, advice, and friendship as well as connecting me with a superb literary agent. That agent, Richard Pine, helped shape the tone of the book and highlight its more general appeal. My editor, Angela von der Lippe, provided the guidance, encouragement, and patience essential to a first-time book author. Erica Stern and Carol Rose offered support and refinements throughout.

I'm indebted to the individuals who graciously provided interviews for the book. Speaking with a group of such interesting and articulate people was one of the highlights of the process. These individuals include: Linda Blade, Ed "Doc" Bradley, John Bramblitt, Brian Bushway, Ben Curry, Zana Devitto, Mike Frueh, JC Ho, Rick Joy, Marty Kaiser, Eileen

Kenny, Daniel Kish, Lisle Leete, Martha McClintock, Marilyn Michaels, Bob Mitchell, Megan O'Rourke, Zoli Osaze, Rick Passek, Jay Patterson, Mark Peltier, Ron Pittman, Steven Poe, Steve Preeg, Dan Radlauer, Dave Reaver, Mark Schmuckler, Trina Shoemaker, Cheryl Tiegs, Sue Thomas, Dave Thorsen, Yasuhisa Toyota, Sacha Van Loo, Chris Williams, and Karl Wuensch. I regret that not every interview could be included in the final version of the book. A special mention should be made of Bob Mitchell who passed away during the book's final editing. Bob had a long, rich, musical life. He loved what he did and people loved him for it.

Thanks also to all those who helped with the photographs used in the book: Alabama Department of Archives and History; Cristin O'Keefe Aptowicz; Art Resource; Artists Rights Society; Glenn Bradie; John Bramblitt; Everett Collection; Institute Nofima Mat; Rick and Wendy Joy; John Ker; John Lykowski; Meredith McLemore; Kjell Merok; Marilyn Michaels; Jacqi Serie; Alison Strum; and Mark, Katalin, and Madeleine Dzinas Schmuckler.

I also thank the following for their feedback, contacts, support, and/or illuminations (listed alphabetically): John Andersen, Gary Bachlund, Dale Barr, Doug Bennett, Lynne Boyarsky, Michelle Brafman, Brew 'n' Beans Coffee House, Curt Burgess, Ben Cabus, Chris Chiarello, Steve Clark, Stuart Contreras, Neal Dykmans, Mike Erickson, Amy Evans, Luis Escalante, Jock Fistick, Robin Fontaine, David Funder, Linda Greene, Peter Hickmott, Ken Kang, Steve Kim, Dai Kim, Cathy Kirscher, Julie Miller, Rachel Miller, Tess Nelson, Lynne Nygaard, Aghadjan Rahbar, Khaleel Razak, Renaissance Kids, Charlotte Reznick, Ryan Robart, Donald Robinson, Rob Rosenblum, Mari Sanchez, Richard Schmidt, Sara Schultz, Aaron Seitz, Deborah Shofstahl, Julie Siguenza, Glenn Stanley, Shannon Tatsuno, and Wendy Wintman.

A special thanks to my family for their support. My beautiful wife, Bita Rahbar, provided important feedback, insight, and patience throughout. And my two young boys, Mateen and Kian, offered joyous, if not relaxing, respite from the writing process.

Finally, I would like to thank all of the perceptual psychologists whose work is discussed in this book, and beyond. We all know that this is the new golden age of perceptual psychology. I hope that this book conveys some of the passion and awe we share for our subject.

See What I'm Saying

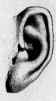

PART I

Hearing

Blind mountain biker Daniel Kish uses echolocation to follow trails and avoid large obstacles.

Chapter 1

The Sounds of Silence

It's a beautiful afternoon in the hills of Mission Viejo. A light breeze provides relief from the high sun and cools my already moist face. The sound of birds mixes with the breeze rushing through the oaks. And as the sun warms their needles, the pines give off their familiar scent.

"Is everyone ready?" Daniel Kish, our guide, asks. "Remember to stay behind each other, but not too close."

Megan O'Rourke, who is new at this, says, "This is kinda scary. But fun!"

"Fun until you crash into me!" Brian Bushway says. We all laugh.

As we leave the safety of Bushway's driveway, we enter the street and hear the plastic pull-ties we've connected to our bike frames clicking against our tire spokes. The sound is very much like that made by the baseball cards kids fasten to their bike wheels to make a faux-motorcycle sound. But today this clicking sound has a very adult purpose.

"Now, Megan, follow the clicking of my wheels so you stay on the side of the road," Kish says.

We turn a corner and I look up at an imposing, apparently endless, upward slope. I think that I wish I were in better shape. I also think that, right now, I am the least fortunate one of our group. I am, after all, the only one who can actually see how much effort we're about to exert. My companions—Daniel Kish, Brian Bushway, and Megan O'Rourke—are blind.

■ ■ ■ ■ ■

I REACH THE TOP OF the hill first, while Kish and Bushway stay back to instruct O'Rourke. As the three of them climb the hill and get closer to me, I start to hear sharp intermittent clicks—different from those of the bike wheels. These sharp clicks are emanating from the mouths of Kish and Bushway, who are using them to hear what I can see. They click with their tongues, about once every two seconds, so that they can hear the sounds reflect back from nearby curbs, shrubs, parked cars, and other obstacles. This method of navigation is known as *echo-location,* and it enables Kish and Bushway to lead these mountain bike excursions. They both click using the side of their tongues, as if coaxing a horse to gallop. And they often change the loudness of these clicks depending on their surroundings. Right now, their clicks sound pretty loud, but they blend nicely with the sounds of the clicking tires and squeaking bikes.

Finally we reach the trail head and begin our official mountain bike ride.

"Here comes the hard part!" I say to Bushway.

"Not for me," he responds. "I prefer riding around rocks, trees, and shrubs rather than the cars, running dogs, and kids in the streets. Mountain biking is more relaxing for me."

And this seems to be reflected in Bushway's echolocating, which is now more sporadic than when we were riding on the street.

As we ride, I ask Bushway, "So what parts of the trail can you perceive from echolocating?"

He responds, "I can hear the sides of the trail where the brush meets the dirt. I can also hear if there are any big rocks or trees in or near the path. All the important stuff about the trail—except maybe the horse droppings. I use another sense for that."

We all laugh.

■ ■ ■ ■ ■

KISH AND BUSHWAY HAVE been leading mountain bike trips for about 10 years. Their Team Bat usually includes 3 to 5 students, but they've led groups as large as 12. All of the participants are severely visually impaired, and most have little, if any, light sensitivity. Besides giving the students a roaring great time, Kish and Bushway believe that these outings build their confidence. They've also been teaching blind students to echolocate, which they feel is one of the most effective means by which they can gain independence. Kish actually conducted scientific research on echolocation for his master's thesis in 1995.

Daniel Kish and Brian Bushway are particularly adept echolocators. Besides mountain biking, they've used echolocation to hike, roller skate, skateboard, and play basketball. Along with his cane, echolocation is Kish's principal way of navigating the world. Kish and Bushway's echolocation skills have also made them celebrities of sorts, landing them on national news and talk shows; in magazine features; and on the lecture circuit, making instructive and inspirational presentations to numerous organizations. In the media, they are often portrayed as "medical mysteries," or as possessing a "special gift." This fact is unfortunate because, while Kish and Bushway's skills are certainly impressive, human echolocation is neither mysterious nor special. And, as you'll soon learn, you too can echolocate, and you do it all the time.

■ ■ ■ ■ ■

DANIEL KISH WAS NEVER ABLE to see. At the age of four months, he was diagnosed with retinoblastoma in both eyes. Retinoblastomas are cancerous tumors of the retina. They are potentially fatal, and treatment often requires complete removal of the affected eye. In Kish's case, one eye was removed when he was 7 months old and the second when he was 13 months. Before then, the blastomas were so large that it is unlikely that he was ever able to visually discriminate anything more than light from dark.

After his eyes were removed, Kish's parents staunchly encouraged his independence. They refused to restrict his activity in any way, despite his potential for bumps and bruises. They also refrained from acting as guides, allowing him to locomote and explore the world on his own. He attributes much of his adult navigation skill, including his expertise with echolocation, to his parents' approach and the confidence it provided him.

Kish remembers himself always echolocating. His parents claim that he started clicking even before his eyes were removed and that he later did it to guide his crawling, cruising, and early walking. By the time he was seven, he used echolocation to bike ride and roller skate. Like most kids, he loved riding his bike around the neighborhood. He was able to hear oncoming traffic and pedestrians by listening for their emitted sounds and stayed safely to the side of the street by echolocating to follow the curb and locate parked cars. Perhaps most impressively, he could judge where he was in the neighborhood by echolocating the driveways between the lawns (hearing the difference in textures) and counting them as he rode. These days, his riding skills are astounding. During production of a television segment about Kish, he was asked to ride his bike around a playground basketball court. The resulting videotape shows him easily riding within the boundaries of the concrete court, and deftly circling the poles that hold up the basket and backboard.

Despite Kish's early expertise with echolocation, until his early adulthood he was unaware of how his clicks were helping him. While he had suspected that sound played a role in his mobility skill, he never experienced his awareness of objects as something he heard, but instead something he "sensed" or "felt." And then in college, he learned about human echolocation in a psychology class. The discovery, which he describes as an "aha" moment, changed his life. After voraciously reading the research literature on human echolocation, he dropped his plan to become a psychotherapist to instead study echolocation and its potential for mobility training.

Kish firmly believes that his knowledge of the research literature made him a much better echolocator, which bodes well for improving your own echolocation skills.

You Hear Silent Objects

THERE HAVE LONG BEEN reports of the blind sensing objects from afar. The eighteenth-century philosopher Denis Diderot wrote of a blind acquaintance who could pinpoint the presence of a silent squirrel successfully enough to hit the poor animal with a stone. Other reports of the blind's ability to sense objects appeared throughout the nineteenth and early twentieth centuries, along with explanations ranging from the blind's presumed sensitivity to magnetism to their clairvoyance. The most prevalent theory held that the blind can sense subtle changes in air pressure on their faces (and other exposed skin) that result from the presence of objects. This "facial vision" theory was based largely on the introspective reports of the blind themselves. But sometimes introspections can be completely wrong about what the perceptual brain is doing.

A definitive test of facial vision was conducted in the 1940s in the laboratory of Karl Dallenbach, in one of Cornell University's old stone-and-wood buildings. Dallenbach's lab was on the top floor and consisted of a large room with a vaulted wood-beam ceiling. Two blind and two sighted men acted as subjects. Each was asked to walk blindfolded toward a large masonite board and stop just before making contact. They repeated this task multiple times and were asked to remain quiet as they walked. All four subjects could perform the task with some accuracy, rarely colliding with the board. When asked how they performed the task, three of the subjects reported feeling changes in air pressure on their face as they approached the board (facial vision). None thought that they were using sound.

But Dallenbach noticed that, though they tried to remain quiet as they walked, the subjects were inadvertently making a great deal of noise. Being that it was the 1940s, the subjects all wore hard-soled shoes. The sound of these shoes against the hardwood floor path was marked. Dallenbach's team considered that despite the subjects' guesses, these sounds might somehow be informing them as to the board's location. To test this, the hardwood floor path was covered with plush carpeting and the subjects were asked to take off their shoes and walk in stocking feet. They also wore headphones that emitted a loud tone to effectively block all incoming sound. Now when the subjects walked toward the wall, they collided with it on *every trial*. Follow-up experiments using methods to neutralize air pressure changes on the skin confirmed that hearing external sound was both necessary and sufficient to perform the task.

After Dallenbach's team established that sound underlies the blind's ability to sense objects, a number of laboratories tested the basis and limits of the skill. Research on bats at the time had shown that these animals use a process of echolocation to determine the position of objects in the dark by emitting high-pitched chirps that reflect off nearby objects and return to their ears. By comparing the time, energy, and frequency differences between the emitted and returning sound, bats are able to determine the location and characteristics of the objects (moths, trees, telephone wires). Dallenbach's team ran another series of experiments to see whether humans echolocated like bats and determined that, indeed, the blind could and did.

Since Dallenbach's work, other laboratories have shown that humans can use echolocation to hear more detailed properties of objects. These properties include an object's horizontal position, relative distance, and relative size. Astonishingly, humans also have the ability to identify the general *shape* of an object (square, triangle, disk), and even an object's *material* composition (wood, metal, cloth) using echolocation. Blind subjects are generally better at echolocating, but untrained sighted subjects are also able to perform all of these tasks with some success and to improve their accuracy with practice. In my own lab's research, we find that sighted subjects can learn to accurately echolocate the position of a moveable board after just 10 minutes of practice.

■   ■   ■   ■   ■

AFTER OUR MOUNTAIN BIKE RIDE, I asked Daniel Kish how he would describe the experience of being an expert echolocator to a sighted person. He provided a lovely analogy: He often camps in the mountains with both blind and sighted friends. His group enjoys late-night hikes and sometimes a sighted friend will bravely forgo a flashlight and let him lead the way. The friend will hold his arm as they walk the trail under the mountain sky and the thick oak canopy that render the path pitch-black. But now and then, the canopy will reveal enough starlight to dimly illuminate the path for a moment. Kish believes that he recognizes these moments by sensing a brief boost in his sighted companion's confidence, which then shows him the frequency and duration of these illuminations. Based on this knowledge, Kish believes that his companion is experiencing something like a visual version of echolocation. Expert echolocation, like night hiking under a thick tree canopy, affords dim "glimpses" of the environment that permit identification of major obstacles and establish the direction to head until the next glimpse comes. Of course, Kish has an advantage over the sighted hiker, because by echolocating at will, he can decide when and how often these glimpses occur.

■ ■ ■ ■ ■

HOW DOES YOUR BRAIN use sound to echolocate? Like bats, you probably use the time delay between an emitted sound and its returning reflected sound to gauge your distance from an object—the farther the object, the longer the delay. The difference in intensity (loudness) between emitted and reflected sound is also likely used for this purpose—the farther the object, the quieter the returning sound.

However, these types of sound cues are limited in their use for two reasons. First, once you are within two meters of an object, your ear is physiologically unable to resolve the very small time and intensity differences between emitted and returning sounds. Second, you don't need to actually emit a sound to "echolocate." I will elaborate on this critical fact later but, for now, it should suffice to say that many echolocation experiments have shown that the skill can be accomplished with sounds emitted by sources other than the echolocator. And, as you'll learn, you can use sound reflections even if you can't *hear* an emitted sound. This makes it unlikely that comparisons between emitted and reflected sounds are necessary.

It is likely that your brain is often using other types of sound cues for echolocation. One of the most important is the sound wave *interference patterns* that occur in front of sound-reflecting objects. The best way to understand this is to try a quick demonstration: Hold your hand up about one foot in front of your face with your palm facing your mouth. Now put your front teeth together, open your lips, and make a continuous *shhhhhh* sound. As you make this sound, slowly bring your hand toward your mouth. You will hear the *shhhh* change systematically as you bring your hand closer. To really hear the sound change, repeatedly move your hand back and forth, closer and farther from your mouth. You will hear a *whooshing* sound that changes with your hand position.

What you're hearing is the sound reflecting from your hand colliding with the sound leaving your mouth. And, as your hand moves, the sound interference patterns change with the distance. You can hear these sound interference patterns change—the whooshing sounds—even if you are not the one emitting the sound. Ask your friend to perform this demonstration near you, and you should have no problem hearing the interference whooshes. If you have no friend nearby, turn on your radio or TV to an unused channel so you hear noise. Move your hand toward the speaker, and again, you should hear the whooshing interference patterns.

While these demonstrations have been designed for you to consciously hear the interference patterns, it is likely that your brain is detecting quieter interference patterns all the time, without your conscious awareness.

■ ■ ■ ■ ■

DANIEL KISH IS NOW the president of the World Organization for the Blind, an association dedicated to helping blind individuals and their sighted friends and family understand the capabilities of the visually impaired. Brian Bushway is also a charter member. A cornerstone of its approach is teaching echolocation for mobility purposes, and educating the general public about its potential. Kish and Bushway travel the world consulting with institutes for the blind and tutoring blind individuals. They have trained over a hundred people to use echolocation with good to excellent success.

What makes one person a better echolocator than another? Kish believes that there are no innate differences that predict a student's success. A student's age when becoming blind can have a small influence on echolocation skill, but it need not. In fact, his best students have ranged from young children with autism to college-aged high achievers to a 65-year-old woman who hasn't been blind very long. Instead, he believes that a student's willingness to use echolocation as a primary means of guidance best predicts his or her success. He says, "Students willing to rely on the skill when they're faced either with a new environment or physically challenging task ultimately fare best with the technique." Put simply, diligent and varied practice is the best way to become an expert echolocator.

Interestingly, Kish's observations are consistent with science's recent understanding of how the brain develops a new perceptual skill. Research shows that the perceptual brain shows a surprising degree of adaptability or *plasticity*. It has long been known that people who lose their sight in childhood will have their brains' "seeing" regions co-opted by the auditory and tactile systems. But even individuals with adult-onset blindness show a co-opting of the visual brain regions by these systems (albeit to a lesser degree than in blind children). The brain changes its structure with even more subtle short-term experience. As you'll learn in Chapter 6, tapping a subject's finger repeatedly in the same location for a *few hours* will reorganize his or her brain to enlarge the area devoted to that finger. It's as if the brain is ready to accommodate even the simplest expertise. Your brain's plasticity holds tremendous potential for your ability to improve your echolocation skills, and to improve all of the skills presented in the book.

■ ■ ■ ■ ■

OF COURSE, TO TRULY understand echolocation, you will need to try it yourself. Here's what I recommend. First, repeat the wave interference demonstration you tried earlier. Close your eyes, make the continuous *shhhhhh* sound, and move your hand toward and then away from your mouth. Now substitute a wall for your hand. Find a wall in your home that has a clear 10-foot path in front and no paintings or tapestries in the area where you'll be echolocating. Position your face about one foot away from the wall, close your eyes, and start making the *shhhhh* sound. Bob your head back and forth so you can once again hear the *whoosh*ing interference patterns. As you continue your *shhhhh*ing, progressively increase your movements so you move back farther before returning to the wall. At some point you will likely not *hear* the wall's presence as much as *feel* it (the classic "facial vision" experience). Do this for a minute or so, and continue to make your movements larger as you do. You should experience the presence of the wall each time you move forward.

Now stop *shhhh*ing, and with your eyes still closed, turn around and walk away from the wall six or seven paces. Turn to face the wall, start making *sh-sh-sh* sounds, and slowly walk toward the wall. Once you sense that you are close to the wall, concentrate hard and move very slowly. When you feel that you're within about a foot of the wall, stop and remove your blindfold. Regardless of how you did, try walking to the wall about 10 more times (without counting footsteps). You'll find that you're quite good by the end of the exercise.

To really improve your echolocation skills, practice this routine every day for a week. It's also a good idea to try it with as many different walls as possible.

How good of an echolocator can you become? It's unlikely that you'll ever be able to mountain bike or play basketball using only echolocation, but with some daily practice and a willingness to try your new skill in novel situations, you'll likely feel more comfortable in the dark and impress your friends with some demonstrations. More important, you'll improve the unconscious and integral echolocation skills you already possess. Recall that you can echolocate based on sounds emitted by a source other than yourself. This type of echolocation, known as *passive echolocation,* is used by all echolocating animals including bats, dolphins, whales, and sea lions. For humans, passive echolocation occurs all the time.

Consider this typical modern scenario. You're late for a morning meeting. As you scamper around hallways on the way to your meeting, you're perusing the notes you hold in one hand, while sipping coffee from a cup you hold in the other. Without averting your eyes from your notes, you adeptly navigate hallway corners, water coolers, recycling boxes, colleagues, and whatever other obstacles face the hallway sprinter. Because your eyes and conscious attention are being largely occupied by other matters, some of your navigation is being directed by sound. And because many of the obstacles you are avoiding make little if any sound, it is likely that you are detecting these things from hearing them *reflect* sound. You are unconsciously echolocating to guide your movements. While your footsteps are responsible for some of the amassed sound reflecting from these obstacles, other hallway sound sources (your colleagues' footsteps and voices, elevator doors opening and closing, even quiet ventilation) contribute even more. In this everyday context, you are guiding your behavior based partly on your ability to passively echolocate.

Certainly emitting one's own sounds for echolocation—*active echolocation*—can help the expert perceive impressive detail. It's unlikely that Daniel Kish would have his astonishing mobility skills without being able to carefully control his tongue's clicks. But for the rest of us, passive echolocation is almost always sufficient. This being said, practicing to improve echolocation by using *active* methods (as in the exercises presented above) could also improve your everyday passive echolocation skills.

You Hear Space

Jay Patterson's job should be easy. But it isn't, and it's mostly your fault.

Patterson is an award-winning sound mixer for film and television. On the surface, his job seems straightforward: he records the sounds of dialogue and action as a scene is being filmed. But his real job is much more challenging. Patterson must capture sound in a way that allows the audience to feel as if they are *in* a scene, rather than in a theater or their living rooms. Audience engagement is paramount in filmmaking. Patterson, whose salt-and-pepper mustache and long ponytail render him every bit the Berkeley native, describes it this way: "The whole idea of a movie is that the audience must be sucked in to feeling as if they are in the scene with the characters. That's a tenuous state to maintain."

You probably have some sense of the effort devoted to making a film set look like a bedroom, an office, or a spaceship, but what you might not know is that substantial resources are put into making the set *sound* like these locations as well. A scene occurring in a bedroom must sound like it for the audience to be truly engaged, and much of the sound of a room is based on its characteristic reflected sound.

Every room reflects sound differently. Try this simple demonstration: Go to your bathroom, stand in your shower, and close your eyes. Now clap once loudly. You should hear a noticeable "echo" sound after the clap. Your shower comprises hard tile or linoleum surfaces surrounding a small area. These characteristics allow sound waves to bounce back and forth, or reverberate, for a longer period of time than they do in many other settings. It is this reverberation that makes your voice sound strong, full, and consistent, which makes the shower an especially inviting place to sing. Now, leave the shower and step into the nearest clothes closet. Shut the door, close your eyes, and clap again. You'll hear that your clap sound is short and muffled and not at all reverberant. There are very few uncovered hard surfaces in your closet, and the soft material of your clothes absorbs much of the sound. If you walk around your house clapping in different rooms, you'll notice that the sound you hear in each depends on its size, surface material, and contents. Perhaps you've noticed that when all the furniture is taken out of a familiar room for purposes of painting or moving, it sounds quite different. In fact, the hollow sound of a newly emptied room often seems to complement the bittersweet experience of moving.

Reflected sound helps your brain recognize the type of space that you occupy. You are rarely conscious of your ability to use sound this way. But when you first enter a space, it is likely that your sense of its general size, shape, and material is based as much on your hearing as on your vision. You are able to instantly perceive yourself in a stairwell partly because your auditory brain detects sound reflections characteristic of stairwells. Your brain is constantly perceiving spaces based on reflected sound, which is one of the ways that you are always echolocating.

And this is how you make Jay Patterson's job so difficult. To keep you engaged in a scene, he must record in a way that closely captures a depicted space's sound reflections. He puts it this way: "Acoustic space sells the reality. If the reflected sound of the space seems discrepant with the visible space, or mistakenly changes during a scene, these glitches have an effect. While they might not be consciously noticeable glitches to the average viewer, they are unconsciously important to everyone and the effect is the same. A glitch in reflected sound will most always throw the audience out of the scene."

Unfortunately for Patterson, recording in a studio or on location poses numerous obstacles to capturing natural, consistent reflected sound. Consider this simple scene:

FADE FROM BLACK.

Angie and Brad are in Angie's bedroom. Ostensibly, she is giving him a tour of her new apartment, but they both know better. Angie is standing at the foot of her king-sized bed, with her back to the bed. She is facing Brad, who is standing just in front of the opposite wall. The wall behind Brad is bare: Angie has not had a chance to hang her pictures. Angie and Brad are about five feet apart, staring into each other's eyes.

BRAD: Why did you bring me here?

ANGIE: I thought you'd want to see the thick red curtains I hung behind my bed.

BRAD: They look fine and they help with the acoustics. But why did you really bring me here?

ANGIE: I wanted to say something . . .

BRAD'S CELL PHONE RINGS LOUDLY. HE LOOKS AT HIS PHONE.

BRAD: It's my wife Jenny. Should I answer it?

CLOSEUP OF ANGIE'S MOTIONLESS FACE.

This simple scene presents a number of reflected sound issues that Patterson must solve. For example, film studios are usually much larger than the sets that are built on them. In this scene, the visible bedroom space is likely much smaller than the cavernous *audible* space of a 50-foot-high studio. Without some type of treatment by the sound crew, Angie and Brad might look like they're in a bedroom but *sound* like they're in an airplane hangar, just the sort of glitch that would throw you out of the scene.

Furthermore, Angie's bedroom, like nearly every room, has different reflections in different locations. The reflected sound near Angie's bed and curtains will be dampened (because of the large, soft structures) relative to the area near the bare wall. This fact creates another problem for Patterson, because the type of microphone typically used to record dialogue is designed to capture sound in a very narrow area and must be aimed at the actor who is talking. Sound mixers become quite adept at turning the boom-suspended microphone back and forth to seamlessly capture dialogue, but the continual swinging of the microphone creates a shift in the recorded reflected sound *behind* each actor. Swinging the microphone between Angie and Brad causes the recorded sound to alternate between the quieter reflections near Angie's bed and the louder reflections of the bare wall behind Brad. This could produce a soundtrack in which Angie sounds like she's in the bedroom, but Brad sounds like he's in a garage. This quick switching of the recorded reflected sound is what Patterson terms "ambient sound shift."

Note that in the real world, ambient sound shift is not a problem. As we listen to two friends chat in a coffee shop, our ears naturally pick up sound reflections from all sides of the room simultaneously, and our auditory sense of the room is determined accordingly. The ambient sound shift problem in film is a by-product of a microphone picking up *only* one side of a room's acoustics at a time, which is something our ears cannot do. Ambient sound shift can create the type of sound glitch that can undermine the audience's engagement with the scene.

So, how does Jay Patterson solve these problems? Whatever he does must be done fast because sound crews usually have less than 20 minutes to resolve all sound issues before filming begins. Here's how Patterson works: He will first walk to the location where the actors will stand during the scene and firmly clap his hands together once. Astonishingly, from this single clap, he can hear all the reflective properties of the set and will know how it must be modified to avoid acoustical problems. He will then instruct his two assistants to hang large, sound-absorbing blankets in precise locations on the set to modify its reflective properties. One blanket might be hung just above the set to silence reflections coming from the full height of the studio. Another might be positioned a few inches in front of a bare wall (behind Brad) to equalize the reflected sound on each side of the room. Typically, four to eight blankets will be hung on a set, all of which must be out of the view of the camera. Correct placement of these blankets molds the room's acoustics to better match the visible space, reduce ambient sound shift, and fix whatever other reflected sound problems the set might pose. Patterson will then confirm that the blankets have been positioned correctly by again producing a single clap. His 20 years of experience have served him well in this regard, because he rarely needs to adjust the blanket positions.

Once he has completed his recordings, the soundtrack is passed on to a postproduction sound team that adds sound effects (the ringing cell phone) and re-dubs the dialogue if necessary. (Brad tends to mumble his lines and often needs to be brought back in the studio to re-dub.) Once again, meticulous attention is paid to reflected sound.

■ ■ ■ ■ ■

SANDY BERMAN IS A SOUND designer who has worked on feature films for 30 years. Sound designers are responsible for fine-tuning or creating all sounds heard in a film besides the music soundtrack. Whether listening to the sound effects of lasers guns and dinosaurs, or simply the clear dialogue heard from characters in an outside location, you are experiencing the work of a sound designer. Much of a sound designer's efforts come during postproduction: after most of the filming and live sound recording have been completed. Berman's job is to add sound effects where necessary, as well as to re-dub any dialogue that has been recorded with less than pristine clarity.

Berman, whose white hair and beard surround a friendly, round face, considers continuity in reflected sound an essential part of his postproduction work. To ensure that all sound effects and re-dubbed dialogue have the same sound characteristics as those captured on the original soundtrack, Berman literally re-creates the acoustic space of the scene in his studio. Using moveable baffles, sound-reflecting walls, and sound-absorbing blankets, Berman custom-makes a recording space to sound just like, say, the bedroom set that had been originally filmed. Amazingly, Berman is able to re-create this acoustic space simply from listening to the reflected sound captured on the original soundtrack. Once set up, all sound effects and re-dubbed dialogue needed for the bedroom scene are recorded in this re-created space. Of course, Berman needs to re-create a different acoustic space for each set in the film.

There are much easier ways to make a recorded sound seem like it's in different rooms, including the electronic echo effects used for popular music recordings, but Berman feels that actually re-creating a live sound-reflecting space is the superior method. "There's nothing like the real thing. While it might seem elaborate, and the audience might not consciously hear a difference, I really believe that the audience can *feel* the difference."

BRAD DECIDES TO ANSWER HIS CELL PHONE.

BRAD: *Hello, Jenny. No . . . No . . . Yes. Goodbye, Jenny.*
ANGIE: *What did she say?*
BRAD: *She knew I was here. She recognized the reflected sound of your bedroom. She's on to us.*
ANGIE: *I guess that's what you get for marrying a fellow acoustician.*
BRAD: *And while I was on the phone with her, I recognized the*

> sound of her *"friend"* Vince's bedroom. He's got a lot of tapestries
> that tend to deaden sound reflections.
> ANGIE: *I love when you talk acoustics.*

Brad closes the bedroom door. They move toward each other.

FADE TO BLACK.

When talking to sound professionals, one does wonder whether their efforts with reflected sound are wasted on the average listener. Perhaps the effort is more for the benefit of the film professionals themselves. But when I suggest this to Patterson and Berman, they each provide a similar response: reflected sound makes a scene's events seem to occur in a real space—a space that the viewer can also "inhabit." Perhaps the most compelling response to this question came from Zoli Osaze, a sound engineer for *Days of Our Lives*, the longest-running soap opera currently on American television (44 years). When I ask whether his audience really cares about reflected sound, he answers, "You might not think so, but soap opera fans are really critical. A few years ago, we changed the baffle positions on the set used for Bo's [a main character] apartment. A week later, we started getting all of these letters from elderly women complaining that Bo's apartment didn't sound right! We had to move the baffles back."

These sound professionals also feel that the audience's ears are getting more discriminating about reflected sound. The general sentiment in the industry is that the playback technologies now available in consumer electronics (CDs, DVDs), as well as movie theater sound systems (THX), have made us all much more careful listeners than we were 25 years ago. This is one reason why Sandy Berman has largely moved from using electronic effects to re-creating live acoustic spaces for his post-production work. "The audience's ear is now much more sophisticated than when I started working. The techniques we once used to create an acoustic space don't work as well for today's audience." So, according to the entertainment industry, not only are you impressively sensitive to reflected sound, your sensitivity continues to improve.

■ ■ ■ ■ ■

Yasuhisa Toyota knows that your ears expect more than they did 25 years ago. This fact has helped make him one of the most successful architectural acousticians in the world. It also makes him nervous.

Toyota and his firm, Nagata Acoustics, have provided the acoustical design of some of the most revered concert halls in the world. These halls include the Suntory and Sapparo halls in Japan, the Mariinsky Opera and Concert Halls in Russia, the renovated Sydney Opera House in Australia, and the Walt Disney Concert Hall in Los Angeles, the city in which Mr. Toyota currently resides. Together with a few other acousticians, Mr. Toyota has ushered in what some consider the new golden age of architectural acoustics. His halls typically have "vineyard" designs: The stages are surrounded by sharply sloped tiers. Each tier is broken into subsections of 100 to 200 seats and these subsections are framed with large side walls, and shorter front and back walls. This design allows for more than 2,000 audience members to each have a clear view of the orchestra, as well as an auditory experience of superb clarity, warmth, and envelopment in the music—all without amplification.

Architectural acousticians typically evaluate the auditory quality of a space along 10 or so different dimensions of sound. The quality of each dimension is influenced by the size, shape, and material composing the space. For concert halls, some of the most important dimensions are *loudness, intimacy, reverberation, clarity, warmth,* and *envelopment.* While each of these dimensions can be defined using physical measurements, experienced acousticians can evaluate a hall simply by hearing the quality of each. Establishing optimal values for each of these dimensions is a goal for concert hall design. But the real challenge is to strike a balance among the dimensions.

Some of the dimensions are noticeable for the inexperienced listener. Most anyone can comment on the *loudness* with which he or she hears an orchestra in a concert hall. Similarly, a listener usually has a sense of how close he or she feels to the source of the music—a dimension that acousticians call *intimacy.* And an inexperienced listener might even be able to hear the *reverberation* in a hall, which, as you've learned, can be understood as the time it takes for a quick, sharp sound (handclap) to stop "echoing" in the space. (Recall your shower and closet.)

Other dimensions are more elusive for the inexperienced listener. Acousticians refer to the *clarity* of a hall as how well each single note can be heard. Imagine an instrumental solo. While inexperienced listeners might be able to make some judgment of clarity, they would likely have difficulty separating it from their experience of reverberation and intimacy. Even more subjectively ephemeral are the dimensions of warmth and envelopment. By *warmth*, acousticians refer roughly to how the hall captures the smoothness of the lower tones produced by the cellos and basses. *Envelopment* refers to the degree to which a hall allows the music to sound as if it is surrounding the listener. Interestingly, good envelopment is based on a listener hearing sound reflections from a nearby surface. This fact helps make the vineyard design, with its multiple side walls surrounding small seating sections, so successful. Not surprisingly, envelopment also adds to the overall intimacy of the listening experience: hearing strong nearby sound reflections makes listeners feel as if they are in a smaller space.

Echoing Sandy Berman, Toyota is acutely aware that modern recording techniques have raised the bar for concert hall design. Most contemporary classical recordings are made by placing microphones very close to each instrument. This provides a recorded clarity for all instruments that would be physically impossible for a single listener to hear in a live setting. "Listeners actually receive a much clearer sound from a compact disc than from a live concert," Toyota says. "They have come to expect clearer sound and this fact has influenced my design process. Clarity must be provided to encourage listeners to come to concerts."

WHAT LESSONS CAN BE taken from the work and skills of sound experts Jay Patterson, Sandy Berman, and Yasuhisa Toyota? Interestingly, all three of them believe that their analytical skills with reflected sound are based largely on years of experience, with only a small part due to some early attunement to sound (they all remember having an early curiosity about sound). They also believe that these skills can be taught. All three have worked with protégés who, under their tutelage, have improved their listening skills.

When I ask Toyota whether he thinks the average listener learns to hear the separate sound dimensions of a hall and discern whether these dimensions are balanced, he offers an illuminating analogy. "People who enjoy wine and drink it often establish a set of subjective scales to evaluate each component of a wine. This allows them to remember and compare wines more easily. Someone who does not drink often, like myself, will not have these scales. The same is true of listening to concert halls. By going to enough concerts in many different halls, the subjective scales will be established."

And when I ask him how orchestras respond to his halls, he laughs. "For the musicians, the first rehearsal in a new hall is a very exciting time. But it's like first driving a new car or moving into a newly renovated house. The steering wheel is a different size, the doorknob is in a different location. Everything is unfamiliar and changes can seem unpleasant. But after some time, you become familiar with the new car or house and learn that many of the changes are for the better. Of course with a new concert hall, the changes are in sound." Toyota has had a number of experiences where, during a first rehearsal, orchestra musicians complain about the acoustics. "They will then come back weeks later and tell me that the sound is now much better and ask what I changed in the hall. I'll tell them that I have changed nothing, and that it is their playing, and *listening* that has changed." It seems that even sensitive listeners can improve how they hear space. And like most of us, these listeners often have little conscious awareness of how space is heard.

Perhaps the more profound lesson to be taken from Patterson, Berman, and Toyota's work concerns *your own* non-expert skills. These sound professionals must develop their listening expertise to accommodate your sensitivity to reflected sound. Speaking for all of us, Jay Patterson says, based on his years of experience, "The unconscious brain is constantly processing reflected sound and every space has a reflected sound 'signature.' Our brains use changes in reflected sound to know that something in the environment is different or that we are in a different space. Reflected sound is integral to our survival."

RECENT SCIENTIFIC RESEARCH ON perception of reflected sound seems to bear this out. Studies have shown that we automatically incorporate the reflected sound properties of a room when determining the location of a sound source (ringing telephone) and that we are sensitive even to the more aesthetic aspects of a room's sound. When listeners are asked to judge the degree to which a musical recording "envelops" or surrounds them, they calibrate their judgments to very subtle changes in a simulated room's properties. By electronically simulating the shift of one wall of a room a few feet, listeners will judge the musical experience as less enveloping. These types of results have informed the concert hall designs of Yasuhisa Toyota and his peers. They also attest to your ability to hear very subtle differences between rooms.

Your ability to hear space from reflected sound is something I've studied in my own lab. My students and I recorded sounds (voices, computer noises, cowbell clangs) in four different rooms: a medium-sized classroom, a small carpeted office, an indoor basketball court, and a bathroom. We then presented these recorded sounds to subjects, asking them to identify the place in which each sound was recorded while looking at photographs of the rooms. Every subject excelled at the task, correctly identifying the rooms 73 percent of the time overall, and the basketball court 93 percent of the time. Follow-up experiments revealed that subjects performed the task by (unconsciously) listening to the decay of the reflections produced in each room and that hearing only a very short portion of the reflection signal was necessary.

So, you can recognize rooms by how they reflect sound. Hearing rooms is another way you echolocate—passively—all the time. This ability is not dependent on hearing reflections of your *own* sounds, but on hearing reflections initiated by other sounds such as voices, music, or footsteps. And there is evidence that your brain actually needs very little sound to perceive spaces. As you'll see, your brain can perceive space based on sounds that are virtually silent.

■   ■   ■   ■   ■

LET'S RETURN TO OUR SCENE with Angie and Brad. This time, however, our scene is being filmed for a soap opera, rather than a feature film. Consequently, the postproduction sound budget is much smaller and Sandy Berman, along with his re-created reflected sound rooms, is too expensive. How can the postproduction sound engineer add the requisite ringing cell phone and Brad's re-dubbed lines so they sound like they exist in Angie's bedroom? Given the constraints, a technique known as "room tone recordings" would likely be used. Room tone recordings are recordings of the quiet, natural sound of the room on a set. Immediately after a scene with dialogue is filmed, everyone on the set is instructed to remain still, while a 60-second audio recording is made of the "silent" room. But in reality, the room isn't completely quiet. The soft sounds of ventilation, humming lights, shuffling clothes, even breathing, produce enough quiet noise reflections to give the room its characteristic sound. As you'd imagine, room tone recordings are so quiet that if you were to listen to one, you'd have trouble consciously identifying any of these sounds.

For our scene, the cell phone ring would be recorded in a quiet sound booth and this recording would be mixed together with the room tone recordings made of Angie's bedroom. This method would provide aural continuity for the scene and be relatively effective (though not as effective as Berman's re-created acoustic space) at making the ring sound like it occurred in Angie's bedroom. Importantly, this method would only be effective if the room tone recordings were made from the exact room in which the original scene was filmed. As Jay Patterson puts it, "The purpose of room tone recordings is to capture the reflecting sound signature of a *particular* room." The fact that the "sound signature" of a room is present in these virtually silent recordings attests to your brain's ability to detect the subtleties of reflections even under the quietest conditions.

There is some scientific research consistent with this conclusion. Daniel Ashmead and his colleagues asked nine blind children to walk down a specially constructed narrow hallway without making contact with the walls. They were asked to remove their shoes and remain completely quiet as they walked. Despite these restrictions, they were able to walk the full length of the hallway multiple times, rarely touching the walls. When Ashmead's team asked them about their strategies, they reported "feeling" or "sensing" the nearness of the walls and then redirecting themselves to avoid collision. None reported using sound to accomplish the task. In fact, the hallway was virtually silent. Still, a subsequent test showed that covering their ears severely degraded their performance, causing them to make frequent contact with the walls. They were originally avoiding the walls by using reflections based on ambient sound so quiet as to be consciously inaudible. They were using "silence."

It seems that the brain is so sensitive to reflected sound that it can use the consciously inaudible sounds of a quiet room to navigate. Certainly blind children and adults will have some initial advantage at this skill. But as you now know, when it comes to perceiving the silent world, practice makes practically perfect.

You Hear Silent Objects That Obstruct Sound

TURN ON A RADIO OR television in your bedroom and set the volume at a moderate level. (The demonstration will work best if you tune your radio or television to an unused frequency so you can only hear noise.) Go into the hallway outside the room and walk about five steps away from the room's doorway. Close your eyes, turn around, and slowly walk back down the hallway toward your room. Stop a few feet from the doorway. With your eyes still closed, face the doorway and see if you can hear the location of the doorframe. As you continue listening, slowly sway your body from side to side to see if it helps you hear the doorframe's edges more vividly. Now, try slowly walking through the doorway and into your bedroom without touching the doorframe.

In this demonstration, your walking is guided by how the silent doorframe covers up or *obstructs* the sounds behind it. This is an example of how you can hear a silent object not by how it reflects sound back to you, but by how it reflects sound *away* from you. The sound waves present in your bedroom were able to move through the open doorway, but they couldn't pass through the rigid doorframe, which reflected much of the sound back into the room. The solid doorframe obstructed the sound, allowing you to "hear" its position.

How good are you at hearing the properties of sound-obstructing objects? The film industry thinks you're terrific. In a filmed scene, whenever a sound is depicted as emanating from behind an obstruction, some postproduction treatment of that sound will be necessary.

FADE FROM BLACK.

Later that evening in Angie's bedroom. Brad is blindfolded. Angie is holding a plate vertically in front of his face. Brad is making click-ing sounds with his tongue. There's a loud knock on the bedroom door.

JENNY (*from behind the door*): Brad, I know you're in there. I can hear you echolocating.

BRAD: And I know you're out there, Jenny; your voice sounds obstructed.

JENNY: Let me in. I have something to say to both of you.

Angie opens the door and Jenny walks in. She leaves the door open.

ANGIE: Before you say anything, Jenny, I need to tell you that it's not what you think. Brad was just showing me his . . .

Just then, loud mariachi music is heard through the bedroom door. Vince, who is in Angie's living room, has switched on her stereo. Jenny kicks the bedroom door closed. The music can still be heard, but it is noticeably quieter and muffled.

ANGIE: Brad was just showing me his impressive sensitivity to reflected sound.

JENNY (*looking sad*): That's a sensitivity of Brad's I haven't seen in years.

BRAD (*surprised*): Oh, Jenny, I didn't know you wanted to see my reflected sound-sensitive side.

Brad and Jenny run to each other and embrace.

FADE TO BLACK.

How would a sound designer like Sandy Berman work on this scene? Typically, for film, all unseen dialogue and nondialogue sound is added in postproduction. For our scene, both the dialogue Jenny spoke from behind the door and the mariachi music would be added in Berman's studio. To render the sound of Jenny's obstructed voice, Berman might have her speak her lines from just outside the re-created acoustic bedroom, and behind some type of small baffle. This baffle would be large enough to muffle the sound of her voice a bit, but small enough so that her lines could be recorded clearly. To add the mariachi music to the scene, Berman might position a loudspeaker playing recorded music outside the acoustic bedroom. Then using a microphone inside the room, the music would first be recorded unobstructed through a door-sized opening. Berman would then close the opening in the room and continue to record from inside the room, as the music emanated from the outside speaker. According to Berman, "Some change in sound must be heard if, in the film, a sound source is situated behind an obstruction. Even if an actor is seen briefly walking behind some type of obstruction, a trash can, for example, in the studio we will record the sound of a crew member walking behind a small baffle."

It might seem somewhat obvious that obstructed sounds do sound different and must be rendered so in film to seem real, but what is less obvious is just how good you are at perceiving obstructed sound. Recent research in my own lab shows that unpracticed, blindfolded subjects can determine a number of characteristics about sound-obstructing objects. Typically in our experiments, thick panels are positioned in front of sound-emitting loudspeakers (typical emitted sounds include soft background speech chatter or quiet static noise), and subjects are asked to judge the characteristics of the panels. Our experiments have shown that unpracticed, blindfolded subjects can determine the horizontal position of two sound-obstructing panels well enough to know whether they can fit between the panels without turning their bodies (the doorway scenario). Subjects can also hear the *vertical* position of a panel well enough to judge if they can walk under the panel's edge without ducking their heads.

Our research shows that your brain can detect even more detailed information about sound-obstructing objects. It turns out that untrained, blindfolded subjects can hear the *shape* of small panels that obstruct loudspeakers. One of our experiments showed that subjects can hear whether a small board is wider than it is tall or taller than wide with impressive precision. And in a particularly astounding feat of auditory perception, untrained, blindfolded subjects showed that they could determine whether a sound-obstructing panel had the shape of a triangle, square, or circular disk. While subjects were not equally good at the task, all were able to recognize the three shapes with accuracy that well exceeded chance performance. Some of our subjects performed superbly, recognizing all three shapes with better than 90 percent accuracy. Interestingly, these latter subjects all reported having some outside experience manipulating sound (as a performing musician, a sound engineer for a church, etc.) suggesting that the skill can be improved with training. But even without training or practice or feedback of any kind (and despite their initial skepticism about the task), blindfolded subjects can recognize whether a silent triangular, square, or circular object is positioned in front of loudspeakers.

Together, our studies indicate something you've likely never realized about yourself: That your brain can detect obstructed sound well enough to hear silent shapes. This is another way in which you are always echo-locating. In these examples, the sound is reflected *away* from your ears, leaving an acoustic hole where the obstruction is positioned. But the principle is the same: your brain uses structured reflected sound to perceive properties of silent objects.

THE RESEARCH, EXPERT SKILLS, and anecdotes presented in this chapter suggest that you do use your ability to perceive silent objects and spaces much of the time. While you might not regularly determine the precise size and shape of objects using reflected sound, it is likely that reflected sound is constantly informing you about the major structures in your immediate surroundings. And while you may not have ever actively echolocated until today, your brain is likely using reflected sound all the time, mostly without your conscious awareness.

As you scurry down the hallways toward your meeting, your eyes are available to read your notes partly because of the wealth of reflected sound your brain is using. As you've learned, you can perceive hallway obstacles by hearing how they both reflect and obstruct sound. You can also perceive when you're coming to a hallway corner by detecting changes in the reflected structure provided by an upcoming side corridor. When you finally arrive at the meeting room, you can hear its doorway well enough to guide yourself through without looking up from your notes. And once you enter the room, you can hear its general dimensions and material properties again without lifting your eyes from your notes. Your ability to consciously multitask is based on an undercurrent of unconscious multitasking accomplished by your perceptual brain.

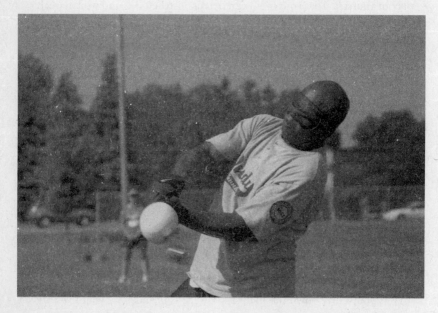

Blind baseball players learn to use the sounds of a beeping ball and buzzing bases for their hitting, running, and fielding.

CHAPTER 2

Perfect Pitches, Beeping Pitches

"C'MON BLAKE, SMACK THE *beep* OUT OF THE BALL!" CALLS ONE OF the Houston Heat players. Blake swings, and hits a searing grounder to left field. He tears off for first base and arrives there safely. His teammates clap and cheer. Bobby, who is about 25 and has a classic ballplayer's physique, is up next. He approaches the plate, slowly swinging his bat to loosen himself. "Close in on the plate, Bobby," his coach suggests. The pitcher takes the ball and . . . "Time out—airplane overhead—can't hear the ball," one of the fielders yells. Members of both teams relax and mill about like kids waiting for a car to pass before continuing their street stickball game.

It's at this point that I'm reminded of just how different beep baseball is. From a distance the game looks like regular softball: A pitcher, catcher, and batter; one team in the field, the other waiting to bat. But from nearby, you notice that there are two large padded bases, the ball emits a pulsing beep, and all the batters and fielders are blind. What little vision these players might have is eliminated by the mandatory use of blindfolds.

The plane passes and the game continues. "Hit that plane, Bobby! Make sure it never comes back," a Houston teammate urges. One of the opposing Austin Blackhawk players in the field claps: "C'mon, guys, let's stop that ball." Another fielder responds, "Phil, you're too close to me, move back a few feet." It seems that the chatter and clapping typically used to keep up enthusiasm have the added purpose in beep baseball of helping players maintain their bearings.

Fonzi, the sighted pitcher, is holding the beeping softball, "Are you ready, Bobby?" Bobby nods. Fonzi draws his arm back past his hip and then calls, "Ready, set, ball" as he throws a firm, accurate, underhand pitch. Bobby hits the ball, at which point the left-side base, located roughly at standard third-base position, starts emitting a loud, sustained buzz. Bobby runs toward the base as the ball bounds down the middle of the outfield. The closer of the two Austin centerfielders dives onto his side to block the oncoming ball. Bobby runs closer to the base and extends his arms to the side to whack the base's large cushioned post, which extends up four feet. But before he arrives, the Austin fielder holds up the ball and yells, "Got it!" Bobby is out.

Beep baseball has been played in the U.S. for over 30 years. There are currently 15 teams across the country, and there is a growing league in Taiwan. The number of games played by each team varies, but the season culminates in all teams congregating for the Beep Baseball World Series in late July. What is most striking about watching beep baseball is not the blindfolds or camaraderie, both of which I expected. Instead, I'm most struck by how seriously the players take the game, and particularly their own performance. As Bobby walks back toward his dugout, he tells his teammates that if he had run straighter toward the base, he would have been safe. A teammate responds, "Next time, Bobby. We'll work on that next practice."

Besides the audible softball (which beeps continually throughout the game) and bases, there are other differences between beep baseball and regular softball. The sighted pitcher is actually on the same team as the batter to whom he's pitching. (The same is true of the sighted catcher.) The pitch is timed and placed with the intent of the batter connecting with the ball (players' batting averages typically run from .500 to .750). This is the reason for the pitcher's "Ready, set, ball" warning, as well as why he pitches from just 20 feet in front of the batter. Each batter gets four strikes rather than three, but each team gets three outs, the one exception being if a fielder actually catches a ball on the fly. This does happen occasionally, and when it does, the batting team's inning is over regardless of how many outs it's accumulated. If a batter gets a hit, he must run to whichever of the two foam-rubber bases is buzzing, which is randomly determined as soon as the ball is hit. The batter must reach the base before one of the opposing team's six fielders has control of the ball, defined as the moment the fielder raises the ball off the ground and away from his body. Three sighted umpires make these and other determinations. Runs are counted whenever a batter makes it safely to a base, so that final scores are often in the teens, despite the game lasting only six innings.

Michael, a 57-year-old with a broad smile, is up next for the Houston team. He has been playing for over 25 years and was part of the group that traveled to Taiwan to teach them the game. Michael laments, "They knew nothing when we got there. But within two years they were beating most of the American teams. It's their speed—those guys are fast."

Fonzi confirms that Michael is ready and calls, "Ready, set, ball" as he throws. Michael hits a pop fly and runs toward "first" base. A few of the fielders yell "pop-up" and the ball drops between the two right-fielders. Michael is running to the base with impressive hustle for a man his age. Just as he hits the base, one of the Austin fielders yells, "Got it!" and holds up the ball. It's close. The members of both teams are silent as they await the verdict. Only the beep of the ball is heard as two umpires confer on the play. Then, after a minute, one of them yells, "Safe!" and the Houston players cheer. An Austin player shouts, "Hey, Ump—ya wanna borrow my glasses? Extra-strong prescription." Everyone laughs.

■　■　■　■　■

HOW ARE THESE BLIND players able to perform so well based largely on the sounds of the ball, bases, and teammates? First, consider batting. Batting is a collaborative effort between the batter, pitcher, and catcher. For each pitch, the batter first takes a few slow practice swings to show the pitcher and catcher where he wants the ball placed. The batter then moves his bat to his back shoulder (normal batter stance) at which point the catcher moves his glove to the batter's selected spot to act as a temporary target for the pitcher. The pitcher then calls, "Ready, set, ball," carefully releasing the ball on the final word. At that same point, the catcher quickly moves his glove back, and the batter starts the process of timing his swing. Using the sounds of the pitcher's call and the beeping ball itself, the batter generally waits about a half second after the ball is released to start his swing. The pitcher and catcher spend days practicing this timing with each of the team's batters, all of whom have different batting styles.

Upon hitting the ball, players must use sound to guide their running toward the base. This skill involves hearing the base's location, as well as anticipating when they will reach that location. Using sound to anticipate their arrival at the base allows them to precisely time when to extend their arms out to the side, making it easier to contact the base's vertical post. If they raise their arms too late, they could miss the base altogether. But, if they raise their arms too early, they could substantially slow down their approach: arm pumping is critical to fast running, so outstretched arms slow the runner.

In a similar way, the blind fielders must use their anticipatory sound skills to time their dive to the ground to stop the approaching ball. If they dive too late, the ball will bound past them. However, if the fielders dive too soon, before they have a good sense of where the ball is going, they could land at an incorrect position outside the ball's path. So again, using sound in an anticipatory manner is a skill that fielders must refine. In fact, some beep baseball fielders seem to have developed an amazing sensitivity in this regard. Michael, the elder statesman of the Houston Heat, believes that from his years of listening to the hit ball, he can hear not only its trajectory, but whether the ball is moving with top spin or bottom spin—features that are informative about how the ball will bounce after hitting the ground.

You Hear the Future

WHAT TYPES OF SOUND information allow the players to accurately anticipate the ball's, and base's, approach? There are a number of acoustic dimensions known to be informative about approaching sounds. As a sound source gets closer to you, more of its acoustic energy reaches your ears. As a result, you hear sounds get louder as they get closer. Relatedly, as sounds get closer, a greater proportion of higher-frequency energy reaches your ears. So not only do closer sources sound louder, they also sound clearer or *brighter*. There are a number of other acoustic dimensions that change with source distance and provide approach information. These include the amount of sound reaching your ears and coming directly from the source, *relative to* the amount of sound reflecting from nearby surfaces; the difference in the sound energy and arrival time at the ear closer versus the ear farther from the source (binaural information); and frequency changes induced by the Doppler shift.

But what is most important about all these acoustic dimensions is the *way they change* as a source gets closer. Assuming a relatively constant sound and speed of approach, all of these acoustic dimensions will change in a way that *provides* anticipatory information. Take the loudness increase as an example. As the ball approaches the fielder, the loudness he hears increases at a rate that is exponential: the loudness increases relatively slowly at first and then increases much more rapidly as the ball gets near. (A result of the inverse square law.) This characteristic modulation of loudness change allows the fielder's auditory system to sample a small portion of the ball's approach sound and determine from that rate of loudness change when the ball will arrive. (Relatively slow change in loudness increase: don't dive yet. Relatively fast change in loudness increase: dive, dive, dive!). Each of the other acoustic dimensions mentioned above (reflections, binaural information) also changes in an informative way as a sound source approaches. These other dimensions also provide anticipatory information.

In fact, this sort of *time-to-arrival* auditory information has been shown to be useful for bats when avoiding obstacles and for pedestrians when avoiding oncoming vehicles. The point to be taken is that the anticipatory timing skills needed for beep baseball might be largely based on players expertly attending to aspects of the changing sound signal itself: a signal that informs them about when the ball (or base) will arrive. The changing sound tells the player something about what will happen a few moments later: the sound allows the player to *hear the future*.

■ ■ ■ ■ ■

WHILE THE HOUSTON HEAT players' skills at hearing the future position of sound sources are impressive, you're not too shabby yourself. You implicitly anticipate the position of sound sources all the time, and your brain seems to be particularly well designed for this skill.

There is an important degree to which hearing is your early warning system. While vision can sample less than half of your surroundings at any instant, hearing samples sounds from all around you. This characteristic of hearing allows it to detect the presence of most nearby things that either make their own sound (a car) or that make sound when they move (which is true of most moving things we encounter). For this reason, you often use your hearing to determine whether some quick reaction is needed: whether it involves moving out of harm's way or turning to look. While the alerting function of the auditory system has long been known, recent research is revealing that the function includes a more sophisticated component designed to specifically perceive *approaching* sounds. While, for most of us, our auditory approach warning system might not have the refinement as those of beep baseball players, it is a system that we are implicitly using all the time.

When you hear an approaching versus receding or stationary sound source, activation in brain regions associated with motion detection, space recognition, attention, and motor reaction is especially high. (A similar neurophysiological reaction occurs with approaching visual stimuli.) The adaptive importance of this system is evident from research on rhesus monkeys showing a similar neurophysiological responsiveness to approaching versus receding sounds. Unsurprisingly, these monkeys also show more agitation when hearing an approaching sound source. Four-month-old human infants display a similar reaction.

The warning purpose of your auditory approach system is revealed in a number of interesting perceptual effects. Research shows that most listeners substantially overanticipate the position of an approaching sound source. In this research, some of which is conducted in my own lab, listeners typically hear recordings of an approaching car whose sound disappears before it reaches them (or more correctly, the microphone used for the recording). Listeners are asked to judge when the car would have reached their position, assuming that it continued toward them with the same speed. Across many experiments with similar methods, all untrained listeners overanticipate the time of arrival of the car, predicting that it will arrive substantially sooner than it actually would have. This pattern of overanticipation reflects the warning aspect of your auditory approach system. The reaction allows for a margin of safety so that in the real world, you will have enough time to avoid harm. It's as if your auditory approach warning system works to fool your reacting brain into acting extra soon, thereby ensuring safety. You hear the future, and it scares you.

But it need not. The auditory approach research shows that with a little practice and performance feedback, you can offset your tendency to overanticipation and become much more accurate. Also, these improvements can be sustained long after performance feedback is no longer available. It is not surprising, then, that visually impaired listeners are generally much more accurate than sighted listeners at the task, often displaying accuracy comparable to that of sighted subjects judging the approach of *visible* objects. It is this plasticity of the auditory approach system that supports the amazing skills of the Houston Heat.

The warning function of your auditory approach system also creates some perceptual illusions. When you hear a sound *increasing* its intensity, its loudness seems to change more rapidly than when the sound *decreases* intensity at the same rate. In other words, sounds seem to change faster when they are getting louder—and, ostensibly, closer—than when they get quieter—and farther. In a related illusion, your perception of a sound source's stationary distance depends on how the source got there. If the source was moving *toward* you before it arrived at that distance, the distance will seem closer than if the source moved away to reach that exact same distance. Again, this supports the warning purpose of your auditory approach system's functioning.

■ ■ ■ ■ ■

THE IMPORTANCE OF PERCEIVING approaching sound sources has recently taken on an unexpected practical consideration that, as of this writing, is garnering significant media and legislative attention.

We all depend on perceiving approaching sound sources to navigate our modern world. But for some, it is a daily matter of life and death. Visually impaired individuals largely depend on hearing the location of automobiles to safely navigate intersections, parking lots, and rural roads. Mobility training for these individuals involves learning to attend to traffic sounds to guide safe intersection crossing, alignment along walkways, and active avoidance in parking lots. For decades, the blind have become expert at guiding their locomotion with the help of vehicle sounds. But now these sounds are disappearing.

There is a modern trend for all vehicles to become quieter so as to reduce noise pollution and the general annoyance of traffic sounds. Certainly, this is a positive change: the widespread negative effects of noise pollution include tinnitus, attentional disorders, and even ischemic heart disease. But as we make the world a quieter place, there is a perceptual cost that needs to be considered. Some vehicles have become *so* quiet that they are functionally inaudible, creating a threat to those dependent on sound for safety. Cars that have alternative power systems such as electric and hybrid models are nearly inaudible when in their electric mode and moving at slow speeds. Undoubtedly, extreme quietness is one of the many appealing features of these new cars. But while quieter cars are certainly a good thing, functionally silent cars might not be. And recent research conducted in my own laboratory suggests that hybrids moving at slow speeds might in fact be functionally silent.

We consistently find that listeners need approaching hybrid cars to be substantially closer than combustion engine cars, before they can hear their direction of approach. Our research shows that this distance depends on the types of cars tested and whether background sounds are present. In the context of a very quiet parking lot, an approaching combustion engine car could be heard at 36 feet from a listener, while a hybrid couldn't be heard until it was only 11 feet away. With the addition of common background sounds (quiet idling cars), the combustion engine car could be heard at 22 feet away, while the hybrid could not be heard until *after it passed* the listener. This last result is particularly troubling.

Our findings confirm the concerns expressed by blind organizations that hybrid and electrical cars moving at speeds below 20 mph pose a hazard to individuals with low vision. But the problem is more far-reaching than this. It is likely that the functional silence of these cars poses a hazard to *all* pedestrians, regardless of their vision. Perhaps you have had the experience of walking in a parking lot, or across a street or driveway, and being surprised by the abrupt appearance of a hybrid car. There are a number of reasons why in the real world, you might not have full visual awareness of nearby cars (talking on a cell phone, managing young children, the physical limitations of age). In these cases, you rely on your auditory approach system to keep you safe. However, the system is only useful if nearby threats are loud enough to engage it. And it is in this sense that the near silence of hybrid cars affects every pedestrian, not just those with visual impairments.

But there is good news. Our research is now showing that the problem can be fixed with the addition of quiet sounds to hybrid and electric vehicles. These sounds will likely be quieter than those of most combustion cars and will not need to contain anything as intrusive as an alarm or warning beep of any kind. Instead, the addition of a very quiet engine-type sound will be enough to make hybrid cars audible from a safe distance. And these subtle sounds will only be required when a hybrid is moving at speeds less than 20 mph and only when it is using its electric motor alone. The solution will require a modification so slight that most pedestrians won't consciously notice the change. It's a solution that will enhance the safety of pedestrians without adding to the problem of noise pollution. And it is a solution made possible by the hypersensitivity of your brain's auditory approach system and its ability to hear the future.

You Hear the Shape of Things

When Sacha Van Loo first heard that the Belgian Federal Police were looking to hire blind officers, he thought it was a joke. But after learning that they were interested in having the blind listen to wiretap recordings, he became intrigued. Van Loo, who has been blind from birth, is fluent in eight languages and has acted as a professional interpreter. He thought these skills might help him decipher wiretapped conversations between suspects that, like Belgium's general population, are increasingly international. He also knew that his blindness made him a thorough listener, a trait also facilitated by his experience as a performing musician. "I don't think my hearing is better than average, I just think that I attend more to sound to compensate for my lack of vision. I thought that this might help the police. Also, I have absolute pitch which I thought might help me recognize voices."

Van Loo joined the force, along with five other visually impaired individuals in 2007. Since then, he has worked mostly at deciphering wiretap recordings of suspected drug and human traffickers. The taps are typically made on cellular phones, which produce recordings that vary widely in quality. "The recordings can sometimes be very poor and for these, listening can be exhausting. But at the same time, deciphering what is happening from a poor-quality wiretap can provide the most satisfaction." Van Loo spends much of his time transcribing conversations between suspects, carefully monitoring who says what. At this, he feels he has an advantage: "I've been told that before we were hired, the sighted police were making many mistakes in determining which suspects made which statements. But recognizing people from voices is something I need to do all the time, and I find this easy even with poor recordings."

In one case, two cell phones with two separate numbers had been registered to a single Farsi-speaking suspect. Both of the phones were tapped, but in listening to the recordings, the sighted officers were unable to determine whether the suspect was using both phones himself or had given one phone to an accomplice. Van Loo was asked to help out. "As soon as I heard the recordings, I knew it was the same suspect using both phones."

Because of Van Loo's experience with language, he is often asked to help identify a particularly exotic language or to recognize the origin of a suspect's accent. In one case, the police were trying to identify a suspect based on his accent. They concluded that he was Moroccan, but checked with Van Loo to be sure. Van Loo listened to the wiretap recording and immediately recognized the accent as Albanian, which turned out to be true.

Van Loo and his colleagues are also able to recognize nonverbal aspects of the recordings. "Depending on the quality of the call, I can often hear where the suspect is calling from: whether he's in a bedroom, or restaurant, or parked car. I can hear the size and type of room behind the voice based on how the sound echoes." He also has the ability to determine what phone number a suspect is calling by listening to the pitch and timbre of the touch tones. One of his colleagues can recognize the type of car heard in a recording based on the sound of its engine.

Despite his impressive listening skills, Van Loo is modest about his success. "We don't solve crimes—our listening skills just speed up the wiretapping process to some degree. And there is nothing unique about our skills. Everyone should be able to do what we're doing with enough practice; it just involves paying close attention to what you're hearing."

And based on recent research on auditory perception, Van Loo may be absolutely correct.

■　■　■　■　■

IN FACT, YOU MAY already have some of the auditory recognition skills needed to help solve a crime, if the situation ever arises. Imagine that you're a rookie detective assigned to investigate a jewelry store heist. There are no obvious clues to the culprit; however, before the heist, the store owner had cleverly placed a hidden microphone in one of the jewel's ring mounts. And lucky for you, the thief is wearing the ring, allowing you to hear everything. Even if your thief is alone and doesn't utter a word, you have the auditory recognition skills to help solve the crime. So, imagine that you put on your headphones and listen for *nonverbal* clues to who and where the thief might be, as well as what he or she is doing and whether he or she is armed.

You first listen for clues about the thief's identity. Based on what is known from research, you would do well to listen to footsteps. Perhaps you've had the experience of recognizing an approaching friend or co-worker simply by hearing that person's footsteps. Certainly, part of this recognition is based on the context (there are a finite number of individuals at your work), as well as the familiarity with the specific sounds of your friends' shoes. Nevertheless, it is still impressive that you can recognize people from the particularly brief, simple sounds produced as they walk. So if you know your thief well—your evil twin, say—you may be able to make an identification based on hearing the footsteps.

But even if you've never met your thief, research suggests you'd be able to determine gender from footsteps. This would be true regardless of the shoes, and if you only heard the thief take eight steps. The research indicates that to recognize gender, you'd be implicitly using the pitch of the footstep sounds to unconsciously apprehend the walker's center of mass (which is higher for men). Related research shows that from listening to the footsteps, you'd also be able to determine if your thief is walking up or down a set of stairs, and whether the thief stoops over when walking, as when carrying the heavy loot.

Continuing your surveillance, if you happen to hear a familiar thief clap hands, you may be able to identify the thief even if you've never heard him or her clap before. Research shows that acquaintances are able to recognize each other's handclaps at better than chance levels, even if they've never previously heard each other clap. Listeners are also quite good at recognizing their own handclaps, which might bode well if the thief is your evil twin. Finally, the research suggests that you'd be outstanding at hearing the *type* of clap that your thief performed: cupped hands versus fingers to palm versus fingers to fingers. (This might help you determine whether your thief is attending a baseball game, golf tournament, or opera performance, respectively.)

Once you've determined characteristics of your thief's identity, you'll want to auditorily determine where the thief is, and what he or she is doing. The research suggests you'd easily recognize nonverbal actions such as cooking breakfast, collating papers, "bathroom chores," and reading in bed. And you'd be able to recognize specific aspects of these events; for example, "The bathroom chores involved hand washing but not tooth brushing; hair combing, but not shaving." In the last chapter, you learned that like Sacha Van Loo, you have the ability to hear the type of room your thief is occupying based on how the room reflects sound. You also have laudable skills at auditorily identifying more specific objects and events including the sounds of specific tools, office supplies, animals, and musical instruments. And if given a list of events to choose from, you'd identify the sounds with near-perfect accuracy. Impressively, your accuracy would not be substantially reduced if you heard the sounds through a severely degraded transmission. This bodes well if your surveillance requires you to listen over a cell phone or even walkie-talkie.

These sound identification findings probably don't come as a complete surprise. You are no doubt aware of identifying simple sounds all the time. What is more surprising is the degree of detail you can hear, even of objects that don't usually make their own sounds.

For instance, you may be able to hear detailed characteristics of your thief's weapon if it happens to be dropped or is struck by another object. Research shows that listeners can determine the length of rods with impressive accuracy simply by hearing them drop on the floor. This ability generalizes to rods with different widths and materials. Analyses of the sound made by the dropped rods suggested that subjects use some complex combination of sound dimensions that result from a mechanical property of the rods themselves: the resistance of each rod to rotational motion (the inertia tensor), which correlates with length.

Not only could you hear a weapon's length, you may be able to determine its rough *shape* simply by hearing it struck. In one experiment, blindfolded listeners heard a series of flat steel plates struck by a metal pendulum. They were asked to identify whether the plates were shaped like a triangle, disk, or square. Amazingly, subjects were terrific at the task, even though they had no training, or prior experience in judging shapes in this way. A second experiment complicated the task by using plates of different materials: steel, wood, and Plexiglas. Despite hearing a much greater range of sounds, listeners were just as accurate at identifying the plate shapes, and were nearly perfect at recognizing which material composed the shapes. The acoustic information for this skill turns out to be complex, likely involving the way each plate produces frequency combinations as well as how sound emanates in different directions and then distinctly dampens over time. Regardless, your hearing brain seems to pick up on the sound signature of a struck object based on its material composition, dimensional details, and even its shape.

So now you've listened to your thief's nonverbal actions and determined gender and identity, what the thief is doing and where (making breakfast, performing bathroom chores); the items the thief's using (pans, soap); and the general size, shape, and material composition of the thief's weapon. You're ready for the arrest. Of course, you'll want to precisely time your entrance and apprehension for when your thief least expects it. If your thief happens to be pouring a drink, you're in luck.

Research suggests that you are excellent at hearing the temporal details of liquid pouring into a container (glass, cup). First off, you are superb at hearing whether the level of liquid in a container is rising, falling, or remaining the same. Also, as you listen to a container being filled from scratch, you are able to hear just when the liquid is reaching the top so that you can avoid overfilling a cup simply from listening. But most impressively, and most germane to your surveillance goals, you can *anticipate* when the container will be full based on hearing just the first few seconds of the liquid being poured. The research suggests that this skill is based on implicitly detecting the changing rate of pitch increase created as the container's empty cavity (and its sound resonance) shrinks to make room for the liquid.

Next time you pour a drink, try closing your eyes and listening for this pitch change. There's a good chance that even without practice, you'll be able to stop your pouring before the liquid overflows. In fact, you'd probably be able to do it even if someone else is doing the pouring. And as you learned earlier, your skills with auditory anticipation go well beyond liquid pouring.

So, by listening to your jewel thief start to fill a beer glass, you could anticipate how long the thief would be distracted by this action. You can now successfully time your entry and apprehension. Congratulations, Rookie, you've solved the case. A promotion is in order and your evil twin is in the clink.

The important lesson is that you hear more than you realize. Not only do you have a general sense of the type of objects that produced a sound, you can recognize substantial detail about those objects and what those objects are doing. This is true even for objects that don't typically make sound. This sometimes becomes evident when you hear a family member drop and break an item in another room: not only do you recognize that something is broken, you often have a sense of what the item is, or at least its rough size, shape, and material composition. But auditory research is now showing that detailed characteristics of objects can be heard in almost any context. As long as there is a sufficient mechanical interaction to put an object into very brief vibration, its general size, shape, and material can often be recognized. And you are likely using this skill all the time, usually without explicit awareness.

You Have Absolute Pitch Memory

THE FIRST THING YOU notice about the Masters of Harmony is the sheer variety of men. From the look of their attire, accountants touch shoulders with surfers; retirees are wedged next to high-school kids. But they are all here for the same purpose: to sing tight, soaring vocal harmonies. The Masters of Harmony is a 150-man, amateur barbershop-style ensemble that has won more national competitions than any similar group in the country. They sing a cappella in an expanded barbershop-quartet style with roughly 40 voices taking each of the four parts usually sung by a single voice. The effect is overwhelming. Hearing 40 bass voices singing loudly in perfect unison provides undeniable chills. But hearing each of the four sections, all 150 voices, sing a harmonized sustained note, feels as if you're being lifted off your chair.

But this takes work. As I watch the director, Mark Hale, take the ensemble through multiple repetitions of the same three-note phrase "Gol-den-gate," it becomes clear that all 150 members are concentrating on singing with absolute precision. The goal it would seem, is to have each section sound like it is *articulating* with a single mouth, but *singing* as if from a giant, organic, bellows.

Hale has the baritones hold the note "gaaaaaate."

"Hear that?!?" he says. "That's the bubble of sound we're looking for. That bubble means we're resonating. That's what we should hear from each section—and *not* vocal strain from any individual singers. Way to go, baritones!"

Cheers and high fives ensue throughout the baritone section.

This "bubble of sound" comes from the singers closely matching their pitch, loudness, and tone (timbre); a notoriously hard thing for 40 men to do, especially in the context of the bouncy barbershop style. But there are techniques that increase the chance of success. First, total attention to the key of the song and the pitch of the individual notes is necessary from each singer. When an a cappella singer's attention to pitch wanders, the muscles controlling the vocal cords become too relaxed, creating a slight lowering of the target pitch. If attention wanders for multiple members, and the ensemble strives to match these fluctuations, a song can erroneously end in a musical key lower than the one in which it began. While this change might not be noticeable to the untrained listener, it is just the sort of mistake that could lose the ensemble a competition.

Tonal or timbral matching across members of an ensemble section is facilitated by having the members produce vowels with the same mouth shape. It is true that the vertically protruding lips seen on singers do help project sound. However, the other purpose of this lip position is for the members to match the timbral dimensions of their notes, creating additional resonances across the overtones of their sung pitch.

A final trick used by the Masters of Harmony to facilitate resonance is to carefully regulate their voice intensity. The goal is for members to sing just under the loudness at which they can hear their own voice as distinct from the ensemble. While being unable to hear one's own voice might seem to undermine pitch and tone matching, with practice it can be the most useful technique to create the desired bubble of sound. Singing below one's own audible intensity makes it less likely that straining voices will be noticeably heard. Modulating loudness in this way also makes it less likely that notes slightly out of pitch with one another will create a *destructive interference* between the sound waves emanating from each singer. This interference dampens the overall intensity of the sound, which in turn can induce singers to sing their slightly off-tune notes louder to compensate. By restraining the amplitude to just below self-audible, singers can defeat this self-perpetuating pitch-loudness error.

From the audience's perspective, hearing the bubble of sound can be a powerful experience. It's as if the emerging sound takes on a life of its own, with the singers simply going along for the ride, with no overt effort. The singers themselves report a similar experience, so it is not surprising that hearing them rehearse can create feelings of awe, but also of jealousy in not being part of the magic. A Masters of Harmony performance is a testament to the potential of the human voice.

■ ■ ■ ■ ■

AND THEN THERE IS your human voice. While you may not have the vocal precision of the Masters of Harmony, you likely sing better than you realize. And despite what you might think from watching inebriated karaoke, and rejected *American Idol* performers, most of us *can* carry a tune when given the appropriate context. In a recent study, individuals were randomly approached in a Montreal public park and asked to sing the Quebec version of "Happy Birthday." These impromptu performances were recorded and then acoustically analyzed in the lab for pitch and tempo errors. The results showed that a large majority of these park goers sung with few tempo errors, and only some slight errors in pitch. In a follow-up experiment, subjects with no musical training were asked to sing the Quebec birthday song at a slower rate. At this slow rate, 85 percent of the subjects produced pitch and rate errors so small as to rival those of a group of *professional* singers who were also recorded (a group that included the actual composer of the Quebec birthday song!). Slower singing likely provides the novice with greater vocal control, including the ability to make quick corrections soon after a note is initiated.

But what of the embarrassing karaoke singer whose vocal prowess is painfully far from that of Celine Dion, or Bono, or even Bob Dylan? It turns out that inexperienced singers are notably worse at *matching* pitches (for example, to instrumental accompaniment), which is just what is needed for a karaoke performance. And what of the *American Idol* hopefuls eliminated in the early, a cappella rounds? Inexperienced singers don't typically have the intensity, range, vibrato, or timbre that professional singers have been trained to enhance. These shortcomings account for much of what we consider an amateur singing style. But still, left to choose your own preferred key and range (without pitch matching), and slowing down your singing just enough to increase vocal control, you can likely carry a tune just fine, often approximating the pitch and tempo precision of professional singers.

And if you do decide to improve your singing by active rehearsal and instruction, chances are that you can make positive changes in your singing, and in your brain. Improving singing can facilitate greater activity in brain areas associated with pitch control and general processing of music. In turn, these brain changes likely enhance the intensity, timbral, and vibrato control characteristic of practiced vocalists. So with some serious practice and guidance, your brain and voice will be ready for karaoke and *American Idol*.

There is, however, a small percentage of people who really shouldn't sing karaoke, at least not in public. In the Quebec birthday song experiments, about 10 percent of the singers made consistent pitch errors, missing nearly every note by a substantial amount. Interestingly, these singers were quite accurate and consistent in their singing *tempo*, as well as in their subsequent *perceptual* judgments of pitch deviations in a piano melody. This suggests that while these singers had a problem with pitch control, their ability to hear pitch deviations allowed them to know how poorly they sang: they were not truly *tone deaf*. True tone deafness, or *amusia*, is a relatively rare trait, affecting about 4 percent of the population. Amusics are deficient in a number of musical skills including the recognition of simple melodies, tapping (and dancing) in time to the tempo of music, and simply determining whether one note is higher or lower in pitch than another. So while you may think of yourself, friend, or spouse, as "tone deaf," it is much more likely that the affliction is one of vocal control, and even that problem affects relatively few of us.

On the other end of the pitch-perceiving continuum from amusics are individuals who can not only determine which of two notes is higher in pitch, but can correctly label the notes (F, A flat) without any context. Roughly 1 in 10,000 adults (in Western cultures) is thought to have "absolute pitch" (the number may be higher in young children). Individuals with absolute pitch, like Sacha Van Loo, can spontaneously name a heard note as easily as they name a color. A number of well-known musicians have either been confirmed or rumored to have absolute pitch (Beethoven, Nat King Cole, and Mariah Carey in the former group; Bach, Ella Fitzgerald, and Jimi Hendrix in the latter); however, it is unclear whether the trait truly confers a musical advantage. There are, in fact, some contexts in which absolute pitch is a small liability. Individuals with absolute pitch are slower to recognize simple melodies that they first heard in a different key ("Happy Birthday" played on the white versus black keys of a piano), a skill all but the amusics of us can perform easily.

While very few of us can label the note of a pitch cold, and it is unlikely that adults can learn to do so, it turns out that a majority of us *are* sensitive to a form of absolute pitch: a form known as pitch memory. Try this simple demonstration. Think of one of your recent favorite songs from your music collection or the radio. Now start singing, humming, whistling, or just vividly imagining the melody of that song. Chances are that you just sang (hummed, imagined) the song in a musical key very close to the key in which the actual song was recorded, and with a very similar tempo. As discovered by Daniel Levitin (and discussed in his provocative book *This Is Your Brain on Music*), when randomly selected subjects are asked to sing or whistle a popular song they know well, a majority of them will reproduce the songs in the same key, or a key within one or two notes of the original recorded songs. These findings are impressive given that there are as many as 11 keys to choose from. And your singing pitch is consistent from one instant to the next even if you're the one who's chosen the key of the song. When mothers are asked to sing their favorite lullaby twice, a week apart, they sing in nearly the same key each time, with pitch deviating *less* than a single note.

Both of these demonstrations show that you have an *implicit* memory for the pitch and key in which a song is performed, whether the performance is yours or your favorite singer's. And even if you were asked to make a more explicit judgment of the pitch of a recorded song, chances are you'd still do well. Imagine yourself as a subject in the following experiment. You're asked to sit in a quiet room, don headphones, and then listen to, of all things, the theme music to "Final Jeopardy" (actually, the first five seconds of the theme). You then hear a few seconds of silence and then the "Jeopardy" theme again, but this time in a slightly different key (shifted electronically), either a single note lower or higher than the key of the first version (for example, a change from C to C-sharp). The experimenter then asks you whether the first or second version of the theme is the correct pitch: the actual pitch that you remember from the TV show. You respond that both versions sound fine and that you don't have absolute pitch, but you provide a guess nonetheless. Research suggests that regardless of your musical training, you'd be correct a majority of the time (as you would with other well-known TV themes like *The Simpsons*, *Friends*, and *Law & Order*). It seems that you are able to unconsciously retain and recognize the pitch of familiar musical excerpts, and do so with impressive accuracy: often within a single note—a margin of error that even individuals with absolute pitch display.

So as you go through your day humming, whistling, singing, or even thinking of your favorite song, you can now have the confidence that you are doing so with quite respectable pitch precision and in a key and tempo very close to that of your favorite musicians.

YOU KNOW THE DEEP STRUCTURE OF MUSIC

YOUR INNER MUSICIAN HAS more sophisticated musical skills. Not only are you good at producing and reproducing individual notes and keys, you know about what happens when notes are combined. Even if you've never taken a lesson, you know something about musical intervals: the pitch difference between two notes. You might not know their names, but you recognize octaves, minor thirds, major thirds, and perfect fourth intervals when you hear them. You can prove this to yourself right now. Sing (hum, whistle, imagine) the first two notes of "Somewhere over the Rainbow." You just sang two notes a full octave apart. Now try singing the first few notes

of "Kumbaya," the old campfire song. The first two notes you sang constitute a major third interval. This is true no matter which pitch you started
with so that if you change the key of the song by singing it in a higher or
lower voice, the first two notes will always constitute a major third. Finally,
sing the first few notes of the ballad "Greensleeves" which you might also
know as "What Child Is This?" (an alternative is the infamous "Smoke on
the Water" guitar riff). The first two notes you sang were a minor third
apart, again regardless of your starting note. Importantly, these last two
intervals, the major and minor third, have particular emotional potency in
Western music. As you'll later learn, major and minor tonality can, respectively, convey happy and sad musical connotations. You may have noticed,
that "Kumbaya" sounds like a happier melody than "Greensleeves."

You can now explicitly recognize some of the most common intervals
used in popular and classical music. The research further suggests that
you would be able to do this regardless of the note each interval started
on: regardless of the key. More importantly, the ease with which you take
to this task suggests that you've likely been *implicitly* recognizing these
intervals all of your listening life.

In fact, the recent music perception research is showing that despite
the fact that most of us are unfamiliar with the technical terminology
of music, nearly all of us have implicit knowledge of these characteristics. Over the last 15 years, the field of music perception research has
changed in two important ways. First, it has implemented brain scanning methods to reveal fascinating things about how the brains of expert
and nonmusicians react to musical and nonmusical sounds. Equally
important, however, is the fact that researchers have successfully implemented methods that allow novices to show off their implicit musical
skills. While more traditional music perception research used subject
tasks that required formal musical training, the recent research uses
more-intuitive, and universal tasks to reveal the novice's implicit sense of
musicality. And this research is revealing that simply listening to music,
as most of us do, can establish these implicit skills. In fact, the more you
listen to music, the more these skills develop.

LET'S CONTINUE TO UNCOVER your implicit musical talents. It seems
that regardless of your musical background, you are sensitive to many

abstract musical qualities. These qualities include how melodic themes relate to variations, how melodies get resolved to sound complete, and how notes in successive chords relate to each other and their implied tonal center (root chord). The take-home message of this research is that *all* of us have implicit knowledge of music's *deep structure*: the complex, dynamic characteristics of music that go beyond simple pitches and rhythms. It is the deep structure that underlies how musical pieces are organized to convey emotion and meaning.

You even have implicit skills with historical compositional styles. You may not explicitly know your romantic from your neoromantic, or your baroque from your classical, but you do know these styles implicitly. When asked to rate the similarity of 16 short piano pieces written in baroque, classical, romantic, and neoromantic styles, nonmusicians rate the pieces of the same historical style as more similar than those from different styles. Nonmusicians also rate pieces from historically *close* styles (romantic and neoromantic or nineteenth and twentieth centuries, respectively) as more similar than those from historically disparate styles (baroque and neoromantic or seventeenth and twentieth centuries).

But before you flaunt your newfound skill, you should know that carp have a similar ability. Actually, carp can discriminate baroque from the blues, but sparrows can discriminate baroque from modern, atonal music, a skill they can generalize across multiple composers of those genres. The fact that musical styles seem so readily discriminable has led researchers to propose that the skill is based on listening for the note-to-note temporal variability in a piece: variability that has generally increased through musical history (baroque music has less variability in the duration of adjacent notes; romantic music has more).

You even have some sophistication with a musical style you may find utterly mysterious, and not all that appealing. Twelve-tone (or "serial") music comprises themes that to many listeners seem atonal, with series of notes often jumping wildly across large pitch ranges. These themes involve notes that appear to have little obvious melodic relation, and sound to many listeners as if they were composed by randomly playing the black and white keys of a piano (all 12 tones in the chromatic musical scale). Twelve-tone music is actually much more sophisticated than this, and its composition is based on a strict rule system, albeit a rule system dramatically different from that used in more traditional Western composition. These rules allow variations to be derived from themes by shifting the theme's notes in key, playing the notes in reverse order, inverting the notes on the musical scale, or both reversing and inverting the notes of the theme. But amazingly, you are implicitly sensitive to this rule system.

In one experiment, novice listeners were asked to listen to a set of 12-tone themes along with a set of variations that were either related or unrelated to those themes. These listeners were able to choose the related variations at levels comparable to those of professional musicians. In a follow-up experiment, novices were able to accurately identify variations that comprised note order reversals of the original 12-tone themes. So, despite the apparent oddity of the musical style, your implicit understanding of 12-tone music runs backward and forward.

Regardless of your experience, you have an implicit sophistication with both the surface and deep structure of music. Of course, none of this would matter if these skills didn't serve your most important facility with music: your ability to grasp its emotional and connotative meaning. As you'll see next, your implicit understanding of these aspects of music is very impressive indeed.

You Know What Music Means

When Bob Mitchell was 12, his mother didn't like him going to the movies. She was a pious woman and, in 1924, respectable children were not seen watching silent films like *The Cigarette Girl of Mosselprom* or *Aelita—Queen of Mars*. Still, the Strand Theatre in Pasadena had the best pipe organ in the area, and Mitchell's mom felt that supporting her son's talent was worth risking his corruption. The young Mitchell's keyboard skills also impressed the Strand's manager, who invited him to play organ between showings of the films. Of course, when the films were actually running, they would be accompanied by a professional organist, a practice that had begun when hiring a small accompanying orchestra became too costly.

On Christmas day, 1924, young Bob Mitchell was playing carols between showings of *The Alaskan* when the film abruptly began. "I found myself accompanying a scene of the film's star, Thomas Meighan, running across a log jam. I quickly started improvising a fast, menacing chord progression on the organ to convey the scene's peril. The next scene was more serene and highlighted the Alaskan terrain. For this, I improvised more melodic, relaxed chords and melody." The theater's professional organist soon returned to take over the accompaniment. But it was clear that Mitchell's talents extended beyond playing Christmas carols: he could improvise music to convey a scene's actions and mood. The 12-year-old Mitchell was immediately hired as one of the Strand's film accompanists.

As Bob Mitchell recalls these events, his blue eyes brighten and a broad smile spreads across his face, belying his 95 years. Mitchell is quite possibly the last of the original silent film organists, a profession that formally ended with the advent of movie sound in the late 1920s. Since the 1930s, Mitchell has directed his renowned "Mitchell Choirboys," a vocal ensemble that has performed in over 100 films, singing with Bing Crosby, Judy Garland, and other luminaries of the period. But even as a nonagenarian, Mitchell still performs organ for silent films biweekly in Los Angeles, and is the genre's most animated and charming historian. Like most film organists, Mitchell largely improvises his scores. Many silent films were actually distributed to theaters with sheet music for accompanying orchestra or organ. But in Mitchell's experience, organists hardly ever used the prepared music, choosing instead to improvise the music scene by scene. Certainly, accompanying the same film for weeks would allow the organist to anticipate a scene's events and moods. But Mitchell claims that he and his peers never played the same score twice.

You are likely aware of how background music can be an important part of the cinematic experience. Film scores can add to a scene's potency by underscoring its actions (think James Bond films), moods (*Casablanca*), and grander themes (*The Godfather*). Film scoring is an undeniable talent that involves painstaking hours of interpreting a film into a recognizable musical language. And this is essentially the task of the silent film organist. However, the film organist has the added challenges of being the sole source of sound during the film, as well as improvising the score live, as he watches the film with the audience. The organist has a unique expertise at spontaneously translating human actions and moods into music.

Sitting with Bob Mitchell at his vintage Hammond organ, I ask how he would accompany a melancholy scene. He immediately plays slow, low minor chords under a descending melodic line. I then ask him to play something for a comedic scene and he quickly switches into a lively melody with a more staccato, bouncy major chord accompaniment. Switching into the romantic passage I've requested, the major chords now slow and become expansive, the melody soars, and I notice Mitchell swaying his body and head as if swooning. As the melody line continues to ascend I hear him say, "And here . . . comes . . . the final . . . kiss!" at which point he plays the resolving major chord while brightening the timbre of the organ with levers. At the same time, Mitchell slowly lowers his shoulders, turns his head, and gives me a glowing smile.

It's at this point that I recognize what makes him such a successful accompanist. It's as if Mitchell takes on the emotion and action of a scene's characters, and then lets his body, hands, and fingers tacitly convey this to the organ's keyboard. By losing himself in the film, his emotional and motoric instincts are able to translate the scenes into music. When I suggest this explanation to him, he agrees and offers an additional and more concrete example. "Here is how I might play a chase scene," he says and starts fingering a very quick repetitive two-note pattern with his left hand and a fast, tight melodic line with his right. I comment that it sounds like the fingers on his left hand are "running" and he responds: "Exactly! And I often use the actual pace of the actors' footsteps for the tempo of what I'm playing." It's clear that Mitchell has a deep understanding of how events and emotions sound musically, as well as how the audience's instincts interpret this musical meaning. The basis of these instincts is something science is beginning to understand.

■   ■   ■   ■   ■

IT's LIKELY THAT YOU'VE always been able to interpret some of the emotional and meaningful aspects of music. Research suggests that you heard music in the womb and responded with physiologic reactions associated with emotion (increases in heart rate and its variability). If, five days after birth, you were played the same music you heard in the womb, you would show similar physiological reactions and would behave more attentively. By the age of four months, you showed clear affective reactions to consonant (melodic) versus dissonant melodies, relaxing when hearing the former and becoming agitated with the latter. (It is speculated that this preference is based on early familiarity with the consonant harmonic structure produced naturally in the vocal tract.)

Research suggests that by the age of four years you could consistently label music based on the emotional categories of happy, sad, scary, and angry. By six, you likely perfected these skills to the degree that you could categorize musical excerpts with the same emotion labels as adults and use the same musical cues to do so. These musical cues are complex, but include the major and minor tonality discussed earlier (major for happy, minor for sad), as well as tempo, with faster tempos interpreted as happier. Interestingly, there is evidence that while emotional associations with major-minor tonality may be learned, the emotional connotations of fast and slow tempo have a more instinctual basis.

As an adult, your emotional interpretations of these cues usually occur implicitly (without awareness of how the cues work) and spontaneously. Research suggests that you can successfully label the emotional connotation of a classical musical passage within hearing its first *quarter second*. And this seems true regardless of your musical background. Brain scans of listeners hearing major versus minor chords show the same patterns in musicians and nonmusicians: minor chords (and dissonant chords) seem to trigger regions associated with "alarm" responses. Support for the automaticity of interpreting musical emotion also comes from research on autistic children. While these children have a very difficult time recognizing emotion from faces or speaking voices, they have near-perfect skills at recognizing emotional connotations from musical passages.

Your impressive skills at identifying emotion in music are likely grounded in your body's physiological reactions. Listening to music produces activity in brain areas known to be involved in emotional responses. Collectively known as the limbic system, these areas are considered some of the brain's most primitive and are involved in regulating reflexive body processes associated with emotional responses. The emotional responses can appear as measurable physiological responses. For example, sad musical passages (from Barber's "Adagio for Strings") can induce large changes in heart rate, blood pressure, and skin conductance. In contrast, happy passages (from Vivaldi's "The Four Seasons") can induce the greatest changes in respiration rates, and "frightening" passages (from Mussorgsky's *Night on Bald Mountain*) can induce the most prominent changes in blood flow rate. Many of these physiological reactions mimic those induced by nonmusical emotions.

Finally, you may not be surprised to learn that when a particularly exciting piece of music gives you a feeling of "chills," your body is likely releasing those much desired endorphins, also secreted by intense exercise, food, sex, and drugs. So endorphins turn out to be the biochemical key linking sex and drugs and rock 'n' roll.

Your implicit emotional reactions to music are probably based on both intrinsic and experiential factors. In fact, your ability to interpret musical emotion might stem from your brain's use of similar sound cues for recognizing vocal emotion. Consider how a voice sounds when it conveys sad news: soft, slow, and low in pitch. Similarly, musical passages that are soft, slow, and low in pitch are typically interpreted as sad. In contrast, voices conveying happiness typically sound louder, faster, and higher in pitch, which are all characteristics used to convey happiness in music. This intrinsic relation between vocal and musical emotion cues might be how very young children interpret musical cues that are common to both Western and non-Western music.

But surely, the basis of your skills with musical emotion also has a critical experiential component. Recall that the emotional associations of major and minor tonality seem based on experience. And certainly many of your emotional reactions to music depend on your own associations (the music of your childhood, the song to which you first fell in love). It's likely that your more nuanced emotional reactions to music such as nostalgia, pride, relief, confusion, and angst are largely based on your personal experience. But regardless of how these interpretative abilities have been established, your implicit skills with musical emotion are spontaneous, automatic, and powerful.

■　■　■　■　■

You are not only masterful at recognizing the emotional connotations of music, you are also a skilled interpreter of more concrete musical connotations, or what is sometimes known as *musical metaphor*. Anyone who's heard Prokofiev's *Peter and the Wolf* is aware of how quickly we recognize musical representations of waddling ducks, stalking cats, and shuffling grandfathers. You may also have an explicit awareness of how musical sounds can convey actions in cartoons and soap operas. But even without explicit awareness, you use more subtle musical connotations whenever you listen to music. The success of some musical styles is partly due to the use of musical metaphor. Consider how the sad wail of a blues guitar or the wry cackle of a klezmer clarinet effectively captures vocal expressiveness. And composers of popular music often use musical metaphor to represent lyrical ideas: the expansive octave interval opening "Somewhere over the Rainbow" so effectively conveys the expansiveness of a rainbow.

Your tacit expertise with musical metaphor is also tapped by film and television composers. Recall organist Bob Mitchell's trick to pace his chase scene accompaniment with the actual footsteps of the actors. I later had a chance to watch him accompany a classic silent film, *The Black Pirate* with Douglas Fairbanks, a film he had never seen, let alone accompanied. It was clear from that performance that his talent to spontaneously translate film action into musical form is extraordinary. Whether musically representing a sword thrust, canon blast, or pirate swinging on sail rigging, his musical metaphors were completely appropriate and recognizable. At various times during *The Black Pirate*, I was able to close my eyes and still have a sense of the actions occurring on the screen simply from listening to Mitchell's accompaniment.

Research suggests that regardless of your musical background, you can easily interpret musical connotation. Much of this research examines how musical meaning is interpreted in the context of visible actions. When watching an animated film portraying a large triangle chasing two smaller shapes, your assignment of different "personalities" to these shapes (playful versus mean) would depend on whether cheerful or menacing music accompanied the film. Similar findings have been reported for more elaborate film scenes depicting the interactions of filmed wolves.

And there is research validating what Bob Mitchell instinctively knows: you understand how music and film action go together. This research suggests that regardless of your musical background, you would have little trouble choosing which professionally scored music goes with which scene in a Hollywood movie. While your experience with the language of film music certainly helps you perceive its connotation, it is likely that the more instinctual bases discussed earlier also play a role.

■　■　■　■　■

So, REGARDLESS OF YOUR musical background, you have implicit skills with the surface structure of music, including its pitch, intervals, and tempo. Your skills also extend to knowing music's deep structure, including its themes, rules, historical styles, and its emotional and metaphoric connotations. The fact that your brain allows these implicit skills to be spontaneous, automatic, and deeply physiological renders music (to quote H.A. Overstreet) the only means whereby we feel emotions in their universality.

PART II

Smelling

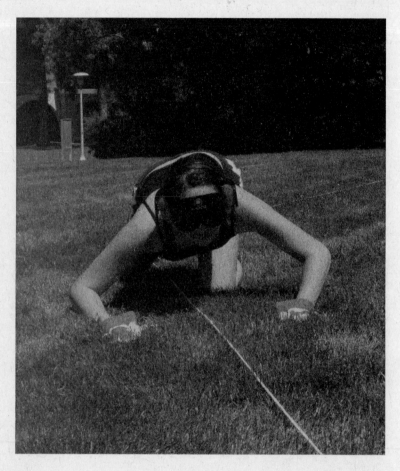

We, like dogs, can track scents using the difference in odor intensity between the two nostrils.

CHAPTER 3

You Smell like a Dog

SOMETIMES TO TRULY UNDERSTAND A SURPRISING PERCEPTUAL phenomenon, you must experience it firsthand. At least this is what I tell myself as I crawl, devoid of sight and sound, across the lawn of my university.

My graduate students and I have decided to try to replicate an experiment demonstrating that humans can successfully track scents, like dogs. My students have blindfolded me, occluded my ears with plugs and industrial ear protectors, and placed thick gardening gloves on my hands. They've spun me around to disorient me and have guided me to the ground to a crawling position. They have also positioned me toward, and about five feet away from, a 40-foot rope they've laid across the ground and secured with garden stakes. This rope has been soaked in peppermint oil for a few days. It is my task to crawl to the rope and then follow its angular path using only my sense of smell.

Once I get over the disorientation from the spinning and absence of vision, hearing, and touch, I concentrate on my scent experience. As I slowly crawl with my nose about four inches from the ground, I get a very strong odor of grass and earth. It's a very familiar and comforting scent, reminiscent of childhood summers. But no peppermint. I lift my head and stick my nose in the air as I've seen dogs do, but it doesn't help. I place my nose back down and continue crawling forward.

Then I get a brief whiff of peppermint. It seems far off and ephemeral, but noticeably spicy, and very different from the earth and grass I've been smelling. I continue forward and the peppermint smell becomes stronger. And then I realize that I've arrived: I am over the rope. I move my head just beyond the point of strongest smell and detect the odor weakening a bit. I move my head back, turn my body parallel with what I believe is the rope, and start crawling along the line of strongest peppermint odor.

As I crawl forward with my nose about four inches above the rope, something interesting happens. I have an almost tangible experience of being inside a shallow trench, or gutter, whose shape is defined by the strength of smell. The bottom-most path along this imagined trench contains the strongest peppermint smell, and the sides are composed of an increasingly weakened smell as they rise from the ground. If my nose moves too far to the side of the trench, it's as if the gradient of smell draws my nose back down the side slope, toward the path of strongest scent—the rope. Perhaps this is what it's truly like to smell like a dog.

But then the scent stops. I move my head to the left and right with no luck. I realize that I must be at one of the corners placed in the rope path by my graduate students to make the task more challenging. I crawl back a few feet, swing my head from side to side and realize that the path angles to the right. I continue to follow the rope, guided by my imagined scent trench, until one of my graduate students taps me on the shoulder indicating that I've reached the end.

I stand up, take off my blindfold and see about 15 undergraduates and a few faculty looking at me and smiling. There is also a student's small dog standing in the grass next to the rope. One of the faculty asks, "What on earth are you doing?"

The dog looks at me and cocks his head.

HAVING TWO NOSTRILS HELPS YOU LOCATE SMELLS

SNIFFING LIKE A DOG puts me in good company not only with dogs but also with some of California's brightest young minds. This human scent-tracking experiment is borrowed from work conducted at UC Berkeley in 2007. The undergrads who performed the tracking task found it relatively easy, if embarrassing. With practice, they were able to substantially improve their scent-tracking skills, often *doubling* their speed at following the trail. Interestingly, to achieve faster tracking, practiced subjects inadvertently found themselves sniffing much more quickly. Faster sniffing is likely part of the superior tracking skills of dogs who typically sniff at rates of six per second compared to our usual one sniff per second.

But perhaps the most interesting finding from the scent-tracking study revealed something astonishing about our noses: we compare smells *across our two nostrils* to determine an odor's location. Have you ever wondered why you have two nostrils? Your nose's symmetry certainly fits the rest of your face. And as you'll learn in Chapters 4 and 8, facial symmetry is an important feature of your attractiveness. But the scent-tracking research has revealed that you also use your two nostrils to determine from which direction a smell emanates.

First, this research showed that subjects have a much more difficult time scent tracking when one of their nostrils is taped shut. Of course this could simply be a results of taking in less air, but a second study using a "nasal prism" showed that when the nostrils get identical input, rather than their normal separate input, scent tracking is again much more difficult. Finally, a particle imaging system revealed that our two nostrils take in air from spatially distinct regions in front of our faces. This might seem surprising given the proximity of our nostrils, but the shape of our nostrils actually serves to funnel in a good amount of distinct air from our right and left sides.

So your nose joins your eyes and ears in making use of two inputs to help you locate where things are. Perhaps your ears are the best comparison. Your brain uses the small differences in when and how much sound reaches each of your ears to perceive a source's location: the ear closer to the source receives the sound slightly sooner and louder than the farther ear. And it's likely that your brain does something similar with your nose—comparing the amount of odor across your nostrils to determine from where an odor originates. (In fact, this location information is sent to a brain area that also merges object location input from the eyes and ears.) Put simply, two nostrils are better than one.

■　■　■　■　■

THE HUMAN SCENT-TRACKING STUDY helps underscore a larger point: you smell much better than you realize. Certainly you don't have the tracking ability, or general smelling skills, of a dog. There are many reasons for a dog's advantage, including the fact that it has roughly 20 times the number of olfactory receptor cells that you do, along with the related genes. And dogs typically have long snouts that are positioned close to the ground and sniff very quickly. But you've compensated impressively for these shortcomings. Your nose's greater distance from the ground has allowed you to go without the complex nasal filtering apparatus needed for dogs, and other ground-sniffing animals, to prevent infection. You may have fewer receptors, but more odor molecules get to them. Also, unlike a dog, you pass odor-infused air over your receptors as you breathe out, as well as in. Most of exhalation scent serves your taste system (Chapter 5), providing you with a much richer range of flavor experiences than a dog has.

But there's no doubt that your most powerful compensatory device for not having a dog nose is your human brain. Smell makes wide-ranging use of your brain, inducing activity in areas associated with emotion, memory, motor reaction, and multimodal integration. Even the language areas of your brain can be directly activated by odors. Activation in these multiple brain areas allows for substantially more processing of the odor input than what occurs in other animals. And the involvement of the language facility allows the educated nose to smell and classify subtleties in wine, perfume, and food.

So what can you do with your nose? Starting with simple detection, it turns out that for some odors, you are actually *better* than dogs. This is especially true of odors composed of long-chain molecular compounds, including ethyl mercaptan, the compound added to natural gas so you can detect a gas leak in your home. Your sensitivity to ethyl mercaptan is so extreme that you could smell it if three drops of the compound were placed into the water of an Olympic-sized swimming pool.

And you can enhance your scent detection skills to notice previously undetectable odors. This is particularly true of androstenone, a compound secreted in human sweat and, as you'll learn, one that might bear on your behavior. It turns out that 50 percent of the population can't consciously detect androstenone, likely because of their genetics. But, with repeated exposure and verbal feedback, even these individuals can "overcome their genes" to detect the compound. And whether learning to detect androstenone for the first time, or simply enhancing your detection of the compound, something astonishing occurs: your nose literally changes. Recent research has revealed that learning to better detect androstenone changes not only neurons in your brain but also neurons in the receptors of your nose. The fact that learning changes the response of neurons in the receptor organ itself is a particularly dramatic form of functional neuroplasticity that, as of this writing, has not been observed for any other sense.

Your skills at *discriminating between* odors are even more impressive. In fact, scientists have found very few molecular changes that can be made to a detectable odor that you wouldn't be able to notice. And for those few odors that are difficult to discriminate, you can usually be taught to notice the difference quickly, albeit sometimes painfully. In one study, two closely related odorous compounds (S-2-butanol and R-2-butanol) were first established to be nondiscriminable for subjects. Both compounds smelled equally "grassy," and both induced the same olfactory brain response. To teach subjects to notice the difference, one of the compounds was then presented along with a mild but unpleasant electric shock on the lower leg. (As nose-related pain goes, not nearly as bad as piercing.) After just seven of these shocks, subjects could then discriminate the two compounds with little trouble. Further, brain scans revealed that smelling the two compounds now induced two different responses in the olfactory brain. Shocking subjects into odor discrimination might seem an extreme way to examine olfactory brain plasticity, but the method nicely demonstrates the power and speed of aversive learning for scent detection, something most animals must contend with in facing potentially fatal smells and tastes.

Fortunately, improvements in smell skills can occur in much more pleasant ways.

■ ■ ■ ■ ■

JC Ho knows the smell of a $65 cup of coffee. He can smell the balance of fruit, cocoa, caramel, nut, smoke, and earthiness inherent to Kopi Luwak, the world's most expensive coffee ($750 a pound). He can also smell its slight tinge of mold, a by-product of how Kopi Luwak is "prepared." The coffee achieves its superior aromatic and flavor balance through the fermentation its beans undergo in the stomachs of Indonesian civets (a cat-sized creature with the facial appearance of an over-caffeinated raccoon). The civets ingest the raw coffee berries, and once in the animals' stomach, the beans go through a process of fermentation. This chemical change is thought to break down the specific proteins that can give coffee a bitter taste. The intact beans are excreted by the civets, picked from the ground, thoroughly cleaned and packaged, and then sent to Ho's Santa Monica coffee shop, along with a handful of other high-end shops around the world. Ho sells about a cup of Kopi Luwak a day and a pound a few times a year.

If you decide to splurge and order a cup of Kopi Luwak, don't bother asking for cream and sugar—Ho won't let you change the flavor in any way. Instead Ho, who is in his midthirties and soft-spoken, will offer to teach you the basics of "cupping," the coffee analogue to formal wine tasting. He'll also show you how he goes about evaluating high-end coffees in determining what to purchase for his shop. This process involves first visually inspecting the roasted beans for color and consistency (his shop does not carry blends), and then chewing on a bean to evaluate its fruitiness and checking for signs of mold.

Then comes the sniffing. "I will finely grind some beans, place the grounds in an empty cup, and inhale the aroma. Evaluating a coffee's scent is as important as evaluating its taste, and I can tell a lot about a coffee from smell alone." From Ho's perspective, a pure, nonblended coffee has five core flavors that can be both tasted and smelled. These flavors are fruitiness, nuttiness, cocoa, caramelization, and earthiness. While a number of the formal coffee organizations employ a flavor wheel consisting of over 100 flavor and smell dimensions (leguminous, horsey), Ho feels that these five dimensions can accurately characterize most any high-end, nonblend coffee (the one exception being Kopi Luwak which, Ho believes, has the extra flavor of "smokiness"). The relative prevalence of each of these dimensions distinguishes, say, Kona from Jamaican Blue Mountain, and Kenyan Double-A from Guatemalan Antigua (assuming similar roasting).

And Ho can smell these dimensions well enough to recognize the coffees, as well as their purity and quality. He also believes that he can recognize the general altitude at which a bean crop has been grown, based on smell alone. "A Kona grown at higher altitudes has stronger earthy and fruity scents. This, I believe, is from the healthier soil at higher altitudes. Sometimes a seller will claim that a Kona is from the mountains, but it won't smell that way. I'll only buy Konas grown in high-altitude soils."

JC Ho wasn't always able to smell this amount of detail in coffees. But just after graduating from college, Ho was traveling through Taiwan where he met his "coffee mentor." "He taught me the cupping process, and how to evaluate the quality of expensive coffees. After much practice I realized that I was able to smell the characteristics of a coffee well enough to recognize particular beans, how they had been roasted, and where they had been grown. A new sensory world had been opened to me."

And this is why Ho would like you to learn about the world of quality coffees. He sees his coffee shop as serving a greater mission: "I want people to enjoy coffee in the same way they enjoy fine wines." And he has some ideas of how to accomplish this using olfactory learning. Ho is currently working on what he calls a "coffee smell sandbox." This collection of six small boxes will include samples of odors which after inhaled will facilitate the recognition of a coffee's unique smells and tastes. A similar technique is sometimes used at wine tastings: a whiff of vanilla can enhance the related flavors in a glass of cabernet.

In fact, there is scientific evidence for a related type of smell enhancement. For example, if you were exposed to high concentrations of a minty smell for three and a half minutes, you would later find other minty smells more discriminable. And your brain would follow suit. Brain imaging reveals that after being exposed to a high-intensity mint compound, areas related to emotion and memory (limbic system) show enhanced activity when sniffing other minty smells. This and other research shows that a relatively short, intense period of exposure to a particular type of smell improves your perceptual and neurophysiological expertise for that class of smells (at least for some period of time). Relatedly, repeated exposure to an odor mixture can enhance your ability to decipher the separate components of the mixture. This is likely the case for JC Ho's expertise with coffee smells, and bodes well for his idea to enhance your expertise using his coffee smell sandbox.

■　■　■　■　■

BUT DESPITE OUR IMPRESSIVE sensitivity at detecting and discriminating odors, there are some aspects of smelling for which most of us are only mediocre. JC Ho notwithstanding, most of us are not particularly skilled at *explicitly naming* specific smells. We often find ourselves faced with tip-of-tongue experiences when trying to name a smell: we can recognize an odor as familiar, and even describe it in some detail (spicy, sweet, burnt), but not quite come up with the correct linguistic label for its identification. (If we are, however, given a choice of labels—citrus, pine—we perform better.)

And even experts can be fooled by an odor's context. In a particularly entertaining example, 54 French wine experts were asked to sniff the same white wine twice, once when presented the wine with its original color and then again after the wine had been secretly tinted with an odorless red food-coloring. Despite smelling the same wine twice in a row, the experts described the red-tinted sample as having characteristics typical of red wines (raspberry, spicy, peppery). This is useful information for the next time you have red wine fans arriving on short notice, but only have white on hand.

And your ability to vividly and consistently *imagine* a smell is even less impressive. Try this demonstration. Close your eyes and try to imagine watching a bag of popcorn pouring into a bowl. Now imagine how that would sound. Most people would describe these "images" as quite vivid. Now imagine the feel of the popcorn as you reach your hand into the popper to pull some out. And try imagining the salty taste and soft mouth-feel of the popcorn as you place it on your tongue. While imagining the touch and taste of the popcorn may be more effortful and less vivid, most people can still imagine these experiences. But now try imagining the smell of the popcorn. For most people, this is very hard to do, and for those who can imagine an odor, the experience usually seems much more transient than images associated with our other senses.

It could be that our failings to both explicitly name and imagine scents have the same basis: we rarely apply complex labels to scents the way we do for sights and sounds, and even things we touch. Of course, this is not true of odor experts like JC Ho who has learned the clusters of scent associated with a coffee's soil, roasting, and native land. And in Chapter 5 you'll learn how applying complex knowledge to a sensory experience can allow professional food tasters and sommeliers to easily label the components of a flavor. But for most of us, there seems to be a disconnect between our impressive abilities to detect, discriminate, and evaluate smells, relative to our mediocre skills at explicitly labeling and imagining odors. Fortunately, most of us are able to effectively use smells at a more implicit level. But not all of us.

You Know the Secret Smell of People

Two near mishaps led Karl Wuensch to seek help for his fading sense of smell. He had just sat down to a leftover casserole he reheated in the microwave. As he was taking his first few bites, his wife came running in and exclaimed, "I can't believe you're eating that putrid stuff!" Together they realized that the casserole had gone bad, and because of its ingredients could have made him quite sick if he had eaten much more. Not long after, Wuensch was working in his yard when he noticed what looked like heat waves rising from the propane regulator on the back of his house. He asked his wife and son to come out and when they did, they immediately smelled a very strong odor of propane. Without the intervention of his family, both of these incidents could have been disastrous for Wuensch.

Upon being tested by his doctor, he learned that he had, in fact, become *anosmic*: he had lost his sense of smell. Wuensch's anosmia was a result of large polyps in his sinuses as well as severely swollen turbinates (the curved bone shelf that guides the air to nasal receptors). He also learned that anosmia can be notoriously difficult to treat, and a majority of the more than 2 million Americans with anosmia go uncured. A scientist by profession, Wuensch thoroughly researched the ailment as well as the compensatory tricks used by anosmics. He installed propane detectors in his house, and took extra care when cooking to watch that pans didn't burn. He also learned tricks to enhance his enjoyment of food, something that had also diminished as he lost his sense of smell. As you'll learn in Chapter 5, it's estimated that as much as 80 percent of flavor comes from smell. Anosmics often enhance their remaining 20 percent of flavor by adding spices as well as experimenting with a food's texture. For Wuensch, hot peppers helped: "I have really learned to love super-spicy food, a love that, sadly, my wife and son don't share."

Fortunately for Wuensch, his anosmia often responds to treatments, at least for some period of time. He estimates that he has lost and regained his sense of smell over four times and that this has given him a unique appreciation for odors. "After a [steroid injection] treatment, I'll drive right home and rummage through my cupboards for smells. I'll pull out crackers, cookies, spices—anything with a distinct odor. And sometimes I'll run outside and bury my nose in the grass or scoop crumbled leaves to my face. It is pure euphoria."

When I ask Wuensch what odors he's been most surprised at missing, he responds: "People. I didn't expect how much I would miss the smell of people. I know of anosmic women who've gone through severe depression when they could no longer smell their child's hair. But I was surprised how much it affected my interactions. It just felt like something was missing in both my intimate and casual interactions. Even though I'd never noticed it before my anosmia, we all have smells, both cosmetic and natural, and it's just not the same being with people when you can't smell them."

■ ■ ■ ■ ■

BASED ON RECENT RESEARCH, Wuensch is absolutely correct. We each have our own special smell, a "scent signature," based on our *natural* odors: not the scents of the soaps, shampoos, and perfumes we don. And while it is not something you typically do at a conscious level, you have laudable skills at recognizing other people's scent signatures.

The research demonstrating this fact typically uses odor stimuli composed of T-shirts worn by individuals who have not showered or used soap of any kind for a period of days. Other experiments use stimuli composed of pads swabbed from unwashed (or, at least, un*soaped*) armpits, and other unsavory areas. Icky as these methods sound, they help ensure that subjects are smelling natural body odors. And lest you think that these methods do not capture how we smell in the modern world, it should be remembered that we only started consistently using perfume and scented soaps a century ago, an eyeblink in human history. The evolution of our bodies, including how our smell systems recognize one another, did not anticipate this recent aesthetic change. Certainly, artificial smells (perfume, soap, shampoo) can provide a compelling association with a person. But tests of person recognition using natural body odors tap into the olfactory skills for which our bodies were designed.

And your skills are impressive. A few days after your birth, you could recognize your mother from the scent produced by the apocrine glands of her underarm and nipple. This ability is based partially on the fact that you could smell in the womb and became familiar with your mother's scent through her contribution to the amniotic fluid. In fact, newborns show a preference for the odor of their own amniotic fluid over those of other children, and parents can recognize the smell of their infant's amniotic fluid. This evidence does not discount the influences of odor learning that also occur in the neonate to an impressive degree.

By the time you were five, you were able to identify the odor of your siblings. This ability has been maintained into your adulthood so that you can still recognize your siblings by smell, despite no longer living with them. And if you have children, you could easily recognize the odor of their T-shirt just a few days after they were born. The one interesting exception is if your children are identical twins, you would be unable to distinguish their scents. This demonstrates the genetic influence on our scent signatures, as the genetic contribution to the odor of identical twins is exactly the same. In fact, parents have much less difficulty distinguishing the odor of their fraternal twins, who have close but nonidentical genetic scent signatures. Strangers also have an easier time distinguishing the odors of fraternal than identical twins even if the twins are no longer cohabitating. And even scent-tracking dogs, who need very little odor to distinguish a tracked person from a foil, are unable to make this distinction if the foil is an identical twin (something to keep in mind if your twin is evil).

Your skills at recognizing your children and siblings from smell also depend on their biological (and genetic) relatedness. While you would have no trouble identifying all your biological children through odor (assuming they weren't identical twins), you might have some trouble identifying your stepchildren, despite cohabitating with both. And while you would be able to recognize the odor of your full biological siblings, you would have some difficulty recognizing your step- or even half-siblings. Besides providing evidence for a genetic component to the human scent signature, these latter findings help illuminate one of the ways that kin odor recognition might work. It could very well be that you recognize your kin's odors through their similarity to your own. While you are likely unaware of having this skill, your implicit, lifelong familiarity with your own natural odor could allow you to recognize odors that have a similar genetic basis: the odors of your genetic relatives.

Still, biological relatedness is not an absolute prerequisite to recognizing a scent signature. Research suggests that you have some ability to recognize the natural odor of friends and have little trouble recognizing your romantic partner from his or her natural odor. This clearly shows how experience can play a critical role in identifying scent signatures.

So what are we smelling when we smell each other? Clearly, there is a strong genetic influence on our natural scent signatures. This influence likely comes from a particular set of genes known as the major histocompatibility complex, or MHC for short. MHC is a particularly dense region of the mammalian genome known to play a critical role in an animal's immune and autoimmune systems, as well as its reproductive success (in giving birth to healthy offspring). Our MHC composition is highly individualistic and is believed to bear strongly on our individual scent signatures. It could very well be that when you recognize someone through scent, either implicitly or explicitly, you are recognizing the biochemical results of the person's MHC genes. An individual's MHC could determine the degree that his or her scent signature is musky, spicy, pungent, or sweet. And as you'll soon learn, MHC can influence your particular scent-signature preferences, mate selection, and even your fidelity to that mate.

■ ■ ■ ■ ■

WHILE THE GENETIC COMPOSITION of MHC is likely the chief influence over your scent signature, there are a myriad of other influences on your odor at any one time. Certainly in Western societies, the colognes, soaps, deodorants, and shampoos individuals choose determine much of their overt scent. (Although there is evidence that one's MHC can determine the *type* of perfume they choose to wear.) These odors can have a conscious, and perhaps unconscious, influence over your recognition and impression of a person. But there are also more subtle, and less intentional, influences on an individual's scent. For instance, a woman's scent will change over the course of her menstrual cycle and is often considered most pleasant during ovulation. Mood can also influence one's odor, and there is evidence that an odor sample taken from a fearful person can be identified as such. Later you'll learn how the odor of someone else's mood can even have unconscious influences over your own *behavior*.

What you ingest also determines how you smell to others. Perhaps you've noticed how particularly pungent foods such as garlic, onion, or curries can influence your odor. And recent research shows that simply eating meat can influence how you smell. A group of men were randomly assigned to either two weeks of eating meat or not. At the end of this period, they each provided an underarm scent sample. These samples were then presented to a group of women asked to rate each sample's pleasantness, sexual attractiveness, and intensity. The results were clear: the odors of the meat-eating group were rated as less pleasant, less sexually attractive, and more intense than those of the non-meat-eating group. Thus, at least for the women tested in this experiment—young women living in Prague—men on a vegetarian diet just smell better: something to consider if you're a single man planning to visit the Czech Republic.

So, just like greeting dogs sharing a friendly sniff for purposes of recognizing each other's identity, family, and recent meals, you can also detect these traits in others, if not with such impressive sensitivity or brazenness.

Like dogs, we also seem to recognize—or at least *appreciate*—each other by what we "leave behind." In a recent study, mothers were asked to rate the scent of solid-waste soiled diapers, including one from their own toddler. Regardless of which diaper the mothers were told was from their toddler, they consistently found the odor of their own child's diaper the least unpleasant. (I have a toddler at home and, frankly, this finding is hard for me to believe. But I defer to the empirical results, hard earned as they are.) The authors of this courageous study explain their results as reflecting an implicit familiarity with one's own child's diaper odor that likely suppresses the usual disgust reaction to a small degree. This familiarity could come from thousands of diaper changes or from the similarity between one's *own* intestinal flora and that of one's child. Future research could examine these competing explanations, but probably won't.

Finally, not only do we smell like dogs, we smell the dogs we *like*. Just as dogs can recognize their owners by scent, owners can reciprocate the love. Research shows that dog owners can nearly always recognize the blanket on which their own dog slept, using smell alone. Furthermore, the owners recognized their own dog's odor regardless of how unpleasant they found it. Dogs, on the other hand, think their owners smell just perfect.

SCENT ADDS EMOTION TO HOW YOU REMEMBER YOUR LIFE

TRY THIS SIMPLE DEMONSTRATION. Go to your bathroom and find a cologne, shampoo, or soap that you haven't smelled for some time. If it is a product that is usually used by another member of your household, that will work even better. Now take the product to a quiet location, open the item, close your eyes, and slowly inhale its scent. Concentrate on what the scent evokes. What does it remind you of? A party from long ago? A romantic dinner? Washing your children's hair and seeing their tiny faces squinch to keep the soap from their eyes? There is a good chance that whatever you're remembering, the memory seems vivid, detailed, and emotionally potent. It was smell after all that induced Proust to write his novel *Remembrance of Things Past* (the smell of a cookie dipped in tea). And the power of smell as a powerful cue for the recall of vivid personal memories has come to be known as the Proustian Hypothesis.

But the truth is, odor doesn't seem to be a more useful tool for retrieving accurate memories than are cues provided by sight or sound. Much of the research supporting this fact has been conducted by Rachel Herz at Brown University and is discussed in her lovely book *Scent of Desire*. But some of the most compelling evidence for this she has published since her book. Imagine yourself in this study. You are first asked to describe an event from your life that is associated with a campfire. In describing this event—a camping trip with your family, for example—you are asked to provide a set of ratings for this memory. You rate how *vivid* and *specific* the memory seems, how *emotional* the memory makes you feel, and the *evocativeness* of the memory: the degree to which the memory makes you feel that you are back experiencing the event.

Once the description is complete, you are then presented with three

types of sensory cues for a campfire: a short, silent video clip of a campfire; an audio clip of the sound of a campfire; and the scent of a campfire
provided through an odor sample (a scented bead). After each one of
these sensory cues is presented, you are asked to think of your camping
trip memory, and again rate its vividness, specificity, emotional quality,
and evocativeness. (The same procedure would then be conducted for
memories related to popcorn and fresh-cut grass.)

Wouldn't you think that the *scent* of the campfire would bring back
the most vivid memories of your camping trip? You'd be wrong. In this
experiment, the visual and auditory cues were just as successful at eliciting vivid, specific campfire memories as the odor cues, based on the
subjects' ratings. While this might be counter to our intuitions about
scent-induced memories, it is completely consistent with previous
research conducted by Herz and others.

What you *would* find different between the sensory-induced memories was their degree of evocativeness and emotional potency. You'd find
that the odor-induced memories of your camping trip seem more emotional and evocative. Note that this would have occurred despite the
fact that your camping trip memory was elicited *before* the odor or other
sensory cues were presented. This is an important point. It means that
the scent cue acted to *render* the memory as more emotional, rather than
eliciting memories that were *already* more emotional. And this finding
that scent can make a memory seem more emotional is consistent with
prior research conducted by Herz and her colleagues.

The finding is also consistent with some fascinating new brain-imaging
results. In a study also conducted by Herz and her colleagues, subjects
were placed in an fMRI scanner and asked to remember personal events
based on the presentation of an odor or visual cue. The odor cue was
the scent of a perfume which the subjects had previously identified as
evoking a pleasant, personal memory. The visual cue was a photograph of
that perfume's bottle. As these cues were presented, subjects were asked
to consider whether the cue evoked a memory and, if so, to think about
that memory. The imaging results were clear: the familiar perfume scent
cue induced greater activity in a core emotional center of the brain—the
amygdala—than the photograph of the perfume's bottle. (This activity
was also greater than that induced by the odor of an unfamiliar perfume.)
These results suggest that familiar scents seem to have a special potency

to induce activity in emotional brain centers, a result consistent with both the perceptual research and our everyday experience.

But as intimated, the research also shows that scent-induced memories are *not* more vivid, specific, or accurate. Why might we, and Proust, experience them as such? The answer could lie with the nature of emotional memories in general. There is substantial evidence that emotional memories are *experienced* as more vivid and accurate, despite rarely being so. Research on eyewitness memory shows that while emotionally charged events often create memories that we *believe* are more vivid and accurate, these memories are typically not more accurate than memories that are not emotionally laden. Emotionally charged memories, whether of personal events (birth of a child, being the victim of a crime) or of more general social significance (*Challenger* explosion, attacks of 9/11, Barack Obama's presidential victory) are often called "flashbulb memories," based on their ostensive retention of the event's vivid detail—as if a flash photograph had been taken. But while remembering an emotionally charged event might evoke emotion during the remembering, the remembered detail is usually comparable to that of nonemotional memories. It seems that the emotionality gives a false sense of confidence in the accuracy of the memory.

Thus, in being able to functionally render a remembered event as more emotional and evocative, odor-induced memories are often experienced as more vivid and accurate than they truly are. Scent induces an illusion of reliving the emotionality of a past event, along with which comes an illusory recall of the factual details of that event. Still, it is a testament to the emotional power of odor, and its limbic system response, that we can be fooled into reliving event details that never happened. Our sense of smell adds emotion to how we remember our lives, an idea with which Marcel Proust would surely have agreed.

You Can Smell Danger

Karl Wuensch's anosmia has been in remission for three years now. He and his doctors have hit on an effective combination of antihistamine, Singulair, and steroid spray. But Wuensch isn't taking anything for granted. "Every morning, the first thing I do is stick my nose in a jar of peanut butter. This turns out to be a litmus test for many anosmics—it's a distinct, strong, and for me, comforting smell." And he feels strongly about people knowing about the plight of the anosmic. "One of the most frustrating things an anosmic contends with is the public's and even the medical community's lack of understanding of how devastating anosmia can be. So many times I've heard: 'You've lost your sense of smell, so what, it's not like losing your vision or hearing.' And it's not like being blind or deaf, but it can still be devastating. Depression is a common theme on the anosmia [Internet] discussion boards. And without being able to smell things that are potentially harmful, there is a hazardous aspect to it as well."

Here, Wuensch has hit on what many olfactory researchers consider the most basic function of smell: the ability to determine whether an odor should be approached or avoided. Despite your failings in explicitly naming and imagining smells, your ability to judge the *pleasantness* of an odor is spontaneous, strong, and automatic. There is evidence that when asked to determine the similarity of odors, humans use pleasantness as the primary dimension. Brain imaging shows that separate brain regions could be distinctly dedicated to detecting pleasant versus unpleasant odors. And three-day-old infants show reactions to odor pleasantness in displaying facial expressions to pleasant and unpleasant odors, similar to those of adults.

These facts have led many researchers to conclude that the most important purpose of conscious olfaction is to evaluate an odor's pleasantness so as to determine whether it should be approached or avoided. The approach/avoidance distinction is the most primitive of sensory processes, evident in the simplest organisms (amoeba). The deep evolutionary basis in evaluating odors could suggest that our sense of an odor's pleasantness would be ingrained and relatively fixed across individuals. But as the Masai tribe in Kenya will tell you, one culture's perfume is another's poop. These tribesman commonly use cow dung to color and scent their hair.

And one's own opinion of a smell can change as our tastes change (beer, cheese) and even after a single event. While you might initially find the smell of a drink or food item fully appealing, a single experience of overimbibing or mild food poisoning can render that smell putrid and intolerable.

These sort of examples have led Herz and other researchers to propose that our odor preferences, while strong, are based largely on our experience and associations with those odors. So, while Americans tend to like the smell of wintergreen, Englanders—especially those born before 1940—generally find the smell strongly unpleasant. This is thought to be a direct result of wintergreen being a component in the analgesic used in Britain during the Second World War. The laboratory research on these *olfactory hedonics* provides other examples. You would likely rate an odor contained in a bottle labeled "parmesan cheese," as much more pleasant than that same odor in a bottle labeled "vomit." The same is true of an odor alternately labeled "Christmas tree" or "disinfectant" or another alternately labeled "cucumber" and "mildew." Verbal labels can also influence your brain's reactivity to an odor, with greater activation of limbic regions when the more appetizing labels ("cheddar cheese" versus "body odor") are presented with the odors.

Herz has also found that a subject's opinion of a novel scent can be changed after just 15 minutes, depending on whether that scent is present when a subject is winning or losing money. It seems clear, then, that your opinion of an odor can be determined by your associations with that odor. And this malleability of odor preferences could make evolutionary sense, as Rachel Herz has argued. Humans tend to live in a wide range of environments, many of which differ in nutritional and hazardous ingestibles. Having an olfactory warning system with flexible preferences could help accommodate the differing and potentially opposing odor cues available across environments.

But other researchers believe that there are important limits to the flexibility of your odor preferences and that these limits have an innate basis. As stated, putrid odors can induce the same reactions in neonates and adults. Also, recent research has revealed common cross-cultural preferences for odors based solely on molecular structure. One study used data from evaluations of over 150 molecules provided by both American subjects and Israeli subjects living in both Arab and Jewish communities. The rated pleasantness of novel odors in all of these communities could be predicted simply on the compactness and heaviness of an odor's molecular structure. This could mean that there is an important innate component to olfactory preferences that appears across individuals and cultures, and is based on an odor's chemical compactness.

In sum, it's likely that your preferences for odors are based on both hard-wired and experiential influences that reflect the plasticity known to exist at many levels of the olfactory system. This constrained plasticity supports the sophistication of the olfactory system in evaluating the pleasantness of odors so as to support your most basic approach/avoidance reactions. They are reactions Karl Wuensch knows all too well.

But while determining the pleasantness of an odor might be the most important *conscious* olfactory process you use, you are doing much more with your nose than you are aware of. And as you'll learn in the next chapter, it's with the odors you don't notice where your nose really shines.

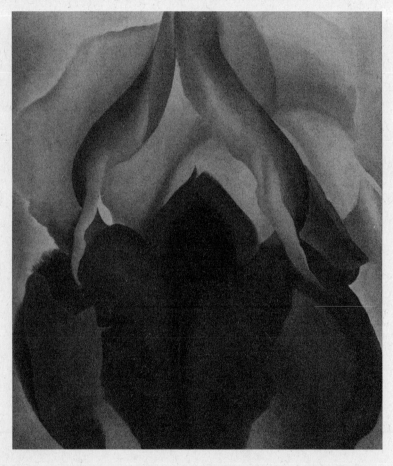

Research suggests that we can unconsciously smell each other's fertility, mood, and genetic compatibility.

CHAPTER 4

Like Marvin Gaye for Your Nose

THE MIRAGE HOTEL IN LAS VEGAS IS WIDELY CONSIDERED THE model for the modern Vegas resort. It was a risky proposition, costing $630 million to build in 1989. But it was a tremendous success, and quickly beat out the Hoover Dam for Nevada's number one tourist attraction. It also became overwhelmingly profitable, eliciting speculations on the reasons for its success, especially from resort developers interested in emulating its appeal. And at the time, one of the most pervasive theories of the Mirage's success was, of all things, its smell. In fact, a distinct odor was intentionally dispersed through the ventilation system of the casino area.

"People were saying that Steve Wynn [the Mirage's developer] had discovered the odor that makes people gamble," Mark Peltier recalls. Peltier's company, Aromasys, was responsible for developing and implementing the aroma dispersed in the Mirage's casino soon after it opened. "I remember how we chose that aroma. Steve Wynn wanted something to match the tropical theme of the hotel. We first came up with some very exotic, sophisticated tropical blends that Wynn didn't like much. And then after a number of months, Eileen [Kenny, Peltier's wife and business partner] said: 'Steve Wynn wants the smell of suntan lotion.' And she was right! We presented him with some aroma blends containing coconut butter and he liked them all. They've been using a coconut butter blend in the casino and lobby for nearly 20 years."

Peltier finds the notion that the smell of coconut butter can make people gamble amusing. "I've spent thousands of hours in Vegas casinos since the early '90s installing our systems and smelling our product. I've never gambled. But I do remember getting calls from Vegas hotels a few years after the Mirage opened. They all wanted coconut butter aroma in their casinos. Even the Excalibur, which has a Medieval England theme!"

Scent has been used in public spaces for thousands of years. The Moors developed a method to preserve frankincense in brick mortar so that its odor would infuse mosques years after building was completed. And it's a well-known realtor trick to bake cookies during an open house to give the home a warm, inviting appeal. But over the last 15 years, the environmental aroma industry has burgeoned, and scents are being dispersed in commercial environments more than ever before. Besides detecting specialized aromas in hotels, spas, and casinos, you're likely to notice them in department and retail stores. Bloomingdale's currently uses a different scent in each of its major departments (baby powder scent in baby apparel; coconut scent in swimwear).

The goals of using environmental aroma are not much different from those of visible décor design. Both methods attempt to make a commercial environment more inviting and relaxing so that customers want to remain in the space. For retail, the longer customers linger, the more likely they are to buy something. For hotels, the more pleasant the environment, the more likely customers are to return. But for Kenny and Peltier, an aroma's pleasantness is not enough: the aroma must fit the space. Kenny designs aromas by considering a number of factors including how a space will be used and the clientele for which the property has been designed. And she has become adept at using scent as metaphor for a lifestyle. "For a more classic, conservative hotel I might include a note of lemon in a blend. For a hotel that caters to a younger, trendier clientele, I might include a note of lemon *rind* instead. For the classic hotel, I might use a lavender note, while in the trendier hotel, I might use a *lavandin*. In general, more conservative clientele prefer the familiar, while the younger, more trendy clientele like things just a touch different." Kenny has also become skilled at matching the components of an aroma blend to the visible décor of a location, a fascinating cross-sensory process you'll learn about in Chapter 11.

Kenny's aromas are then carefully dispersed through a building's ventilation system using a proprietary electrostatic vaporization system designed by Peltier. This system has worked well for Kenny and Peltier: Aromasys services many of the country's high-end hotel chains. Yet, as Peltier comments, "We feel *most* successful when we hear that customers are providing unsolicited positive comments about the scent in a lobby. Our goal is really that simple."

But both Kenny and Peltier have heard questions about the assumed coercive power of environmental aromas. In the early 1990s, Peltier dealt with the issue frequently. "Katie Couric burst into our offices with cameras and bright lights and started interrogating me. She asked how we manipulate hotel and retail customers with our aromas. I told her that manipulation is not what we do. Our system just adds to the décor of a site, similar to the color of walls and background music. And we use aromas that are no different than those found in nature. No one would accuse a forest or flower garden of manipulating us into doing things we didn't want to do. We just make the commercial properties more pleasant places."

But Peltier is aware that aroma is considered a more mysterious, and to some, more threatening, décor component than wall color. "It goes back to ancient times. Aroma is the 'unseen force.' It's ephemeral, fleeting, and unpredictable. And it has a direct connection to our emotions. These things aren't usually true of what we see and hear."

■　■　■　■　■

So SHOULD YOU BE SCARED of what you smell? Can odor influence you to make decisions you'd rather not make, and in more potent ways than the effects of what you see and hear? Probably not. Marketing researchers have shown that people will rate a store (or laboratory simulation of a store) as more positive if it is infused with a pleasant scent, and say that they are more likely to buy something. Other studies show that patrons will spend more time in a pleasantly scented store or casino and will be more likely to use a slot machine that is in a pleasantly scented area of a casino. But the research also shows that these same effects can be induced by pleasant *looking* décor or nice *sounding* background music. And while a store's scent can, in theory, render a patron's thoughts and memories as more emotional, it is also more likely to induce a strong negative reaction in patrons if they find it unpleasant (as you learned in the previous chapter).

So while environmental scent may be a novel addition to a hotel or store's décor, there is no evidence that it has special coercive powers. A smell can't brainwash you into gambling, or staying at a hotel you don't like, or buying something you don't need.

This being said, new research is revealing that your nose does actually do something sneaky; and it's what you *don't* smell that can fool you.

You Use Subliminal Odors

THE RECENT SMELL RESEARCH suggests that you are being affected by odors you never notice—possibly more often, and more strongly, than by the ones you do. Your everyday experience is that your nose spends much of its time "on-hold," waiting for the next novel odor to waft by. When entering a new environment you do notice new odors, but once habituated, your nose seems to be at rest. Unlike your eyes and ears, which always seem to be detecting something, much of your waking time seems scentless.

But new research suggests that conscious smelling is just a very small fraction of what your nose does. Your nose is likely using odors all of the time, mostly in ways that don't reach your consciousness. New research in physiology and neuropsychology reveals an olfactory system that is especially suited to *subliminal* input and processes. What's meant by *subliminal* here (and by smell researchers) is simply when an odor is too weak to reach the threshold of your conscious awareness: it is unnoticeable. Importantly, subliminal does *not* itself imply coercion or manipulation of any kind.

Despite the fact that they are too weak to be noticed, subliminal odors activate the nerve cells in your nose and induce activity in brain regions associated with your attention, memory, and emotion. This activation is similar to that initiated by consciously detectable odors. Even when you do consciously notice an odor, unconscious odors still influence your brain. In fact, by the time you notice a smell, your brain has already been activated by the smell a half second earlier: a lag that is much longer than for any of your other senses. And even after the odor reaches your consciousness, activity in regions specifically involved with subconscious odor processing continues to run in parallel with conscious odor brain processes.

This new evidence has led a number of researchers to believe that much of your olfactory processing occurs on stimuli that never reach your awareness. And this hidden olfactory world can have subtle effects on your thoughts, preferences, and to some degree your behaviors.

Relevant to the environmental aroma industry, there is evidence that associations between a space and its smell can occur at a subconscious level. Imagine sitting in a room for 45 minutes and performing a series of math and creativity tasks. Unbeknownst to you, a very low level of lavender odor is pumped into the room as you work. The odor level is so low that you have no conscious awareness of its presence. You are then brought to a new room and asked to, essentially, play the part of a scent designer like Eileen Kenny. You're asked to look at photos of 12 different rooms as you smell 12 different detectable odors. Your job is to rate how well each odor fits with each room. One of the photographs is of the test room in which you just sat, and one of the odors is the same lavender that was pumped into that room at undetectable levels. Chances are that you would choose the lavender odor as being the best fit for the test room. And because you were unaware of the odor in the original room, the association of the room to odor occurred subliminally.

Other research shows that subliminal odors can subtly influence very *specific* thoughts and even behaviors. For example, you would more quickly recognize cleaning-related words ("hygiene," "tidying") on a computer screen if a low-level, subliminal cleanser odor were presented as you responded. Also, this subliminal cleanser odor could influence what you reportedly planned for your day. Subjects are much more likely to report plans for cleaning-related activities (straightening their rooms, washing their cars), in the presence of an undetectable cleanser odor. Finally, a subliminal cleanser odor can actually help you become more tidy. Subjects who are asked to eat a crumbly cracker perform three times as many cleaning actions (wiping crumbs from their desk and mouth) if they are first exposed to the subliminal cleanser odor.

Not only can odors influence you unconsciously, odors can influence you when you're *unconscious*. It turns out that you sniff in your sleep and what you sniff can influence how you dream. Research shows that presentation of a rotten egg odor induces sleepers to have dreams of more negative emotional content, while presentation of a rose odor brings on more positive dream emotions. Sniffing in your sleep can also influence how you remember things upon waking. It's long been known that the *slow wave* phase of sleep helps you consolidate your memories of the day. If those memories (of a computer task) are established in the presence of a rose odor, later presentation of that rose odor during slow wave sleep will help you recall the task better upon waking the next morning.

In sum, research shows that subliminal odors can have subtle influences over your learning, planning, and behavior. And these influences can occur even when you're sleeping. Of course the types of odors bear on the influences that can occur. And some of the most important types of odors are those of people.

You Can Smell Fear

HAVE YOU EVER TAKEN an instant dislike to someone, but weren't quite sure why? Perhaps your nose knows. It could very well be that your impressions of people are formed partly by how you unconsciously find their odor. Research shows that the presence of an undetectable odor can influence the degree to which you find a face likable. In these experiments, if the subliminal background odor is a low-level lemon scent, subjects typically rate presented faces as more likable than if the background odor is neutral (anise) or unpleasant (sweat). Interestingly, these effects occur *only* when the background odors are unnoticeable. Once the odors are strong enough to be noticed, no influences on the faces' likability occur. Perhaps once the odors are noticeable, subjects are able to consciously resist what they assume to be the expected influence. The next time you get a negative impression of someone without knowing why, perhaps it would help to get a stronger, more noticeable whiff of him or her.

But don't forget that you have an odor too. And while constant exposure to your own odor makes it difficult to notice, it's likely you are subliminally perceiving your own smell much of the time. Research has shown that subjects are significantly faster at recognizing images of their own face when being presented subliminal amounts of their own odor. It's as if subliminal amounts of your odor are always available to remind your unconscious mind of who you are and how you smell. And as you learned in the previous chapter, knowing your own odor can help you recognize your genetic kin through their scent.

Your unconscious smell skills also allow you to pick up on someone else's fear. If you were exposed to low levels of odor samples taken from nervous men before they entered an exam, your interpretation of facial expressions would be more negative. These same odor samples would also make you a little more "jumpy." Your *startle reflex* is the automatic eyeblink response you have when you hear a loud, abrupt sound (popping balloon, uncorking champagne bottle). It is a well-studied reflex known to be strongly influenced by one's emotional state. It turns out that the startle reflex is amplified when subjects are presented a small amount of fear-induced odor. You unconsciously know the smell of fear, and it makes you nervous.

But a little fear can be a good thing. Exposure to a fear-induced odor can actually improve your performance on some tasks. For example, when subjects are asked to determine whether pairs of words are related in meaning (HAT, CAP) or not (LEG, CUP) their accuracy increases when they are exposed to low levels of a fear-induced odor. Fear odors might put you in a heightened state of cautiousness and vigilance, making you more careful on some tasks.

The influences of fear odors are well known in the animal world. Animals ranging from amoeba to deer can communicate fear through odor, inducing behavioral, physiological, and even immunological changes in the odor recipient. Ants, for example, immediately disperse when they detect the odor of another ant under duress. Mice warn one another of predator location by leaving fear-induced odors wherever they've been threatened. Given the prevalence of fearful-odor communication in the animal world, it is conceivable that we have a similar system, or at least its remnants.

You Know the Odor of Fertility

People didn't discuss menstruation in public back in 1969. That's why a roomful of male biologists turned to stare at young Martha

McClintock when she brought up the topic. At the time, McClintock was a college junior enrolled as a summer intern at a prestigious New England research institute. She was responsible for the "graveyard shift" in one of the institute's mice labs. Despite her constant exhaustion, she attended a weekly seminar made up mostly of renown biologists. On that day, the biologists were discussing how the olfactory-based, chemical messengers known as *pheromones* induce cohabitating female mice to synchronize their reproductive cycles. McClintock recalls the experience well: "I was the only woman, and the only undergraduate in the room. I remember feeling deeply embarrassed and blushing as I raised my hand. I made the point that women living together also have synchronized menstrual cycles. I then remember all of these eyes staring at me. One of the men asked what I meant and whether I had proof. That made the situation even worse."

McClintock remembers later standing outside the seminar room. "A male postdoctoral student was teasing me for my comment that women synchronize. I said, 'But it's true!' So we decided to make a bet." And it was a bet well-made. Not only did McClintock win, but the bet started one of the most important, and controversial, research areas in human olfaction: the study of human pheromones.

In truth, McClintock had inside information. She was a student at Wellesley College where she lived in a dorm with over 100 women. She was the member of her dorm responsible for riding her bike to the drugstore to pick up supplies. She noticed that for most of the month, there would be an ample supply of feminine hygiene products. "But then once a month people would start yelling, 'We're out!' and I would get on my bike and ride down to the drugstore. After observing this for many months, I realized that the women were having their periods at about the same time." This motivated her comment in the biology seminar on that embarrassing summer day.

Upon finishing her summer internship, McClintock returned to Wellesley to win her bet. Along with her adviser, she conducted formal research on menstrual synchronization of women. She enlisted the help of her 135 dorm mates and interviewed them every few months. She asked the women about the regularity and dates of their menstrual cycles, as well as with which other women they had the most consistent contact (friends, roommates). She also asked how much time they spent in the company of men each week. During the interviews, she would talk to the women about the physiology of menstruation and the fertility cycle. "I was recently back at Wellesley to give a talk, and there was a woman in the audience who had actually been in the study 40 years ago. She stood up and said how much she appreciated me teaching her about menstruation. It was just not something that was discussed openly back then—not even between mothers and daughters."

The data obtained from McClintock's interviews confirmed her casual observations. When the women first moved into the dorm, their menstruation occurred at different times of the month. After four months of cohabitation, however, the women's cycles shifted so that they were roughly in sync with those of the women with whom they had the most consistent contact.

In her original 1971 report of the study, McClintock speculated that the basis of the menstrual synchronization was similar to that of other mammals. As she had learned during her summer internship, mice and some other mammals menstrually synchronize through their detection of the pheromones emitted by nest mates. The chemical composition of pheromones changes across the reproductive cycle. These changes can be detected by an animal's olfactory system, which, in turn, can shift the reproductive cycle of that animal. This accounts for the reproductive synchrony that has been observed in mice and some other mammals. McClintock thought that synchronization in women occurs the same way: repeated exposure to other women's pheromones could induce menstrual entrainment across the women, even without their awareness.

But McClintock's findings, and the whole notion of human pheromones, are not without controversy. Regarding the former, it turns out that self-reports of menstrual cycle dates are often inaccurate. It could be that the women's memories of their own menstrual dates were influenced by what they remembered about the menstrual dates of their roommates and friends. In fact, half of the women in McClintock's study did report having some idea of when their friends' menstruation occurred. A few subsequent studies have not replicated McClintock's results, possibly for this very reason. Next, a replication that did find menstrual entrainment found evidence that synchrony is most likely to occur between women who are close friends regardless of whether they live in the same room or even on the same floor. This seems odd if menstrual synchrony is based on olfactory sensitivity to pheromones alone.

But throughout her career, McClintock has provided other evidence for putative pheromone influences on menstruation, mood, and arousal. After graduating from Wellesley, she went on for her doctorate partly under the direction of E. O. Wilson, the renowned insect pheromone researcher and the father of sociobiology. She then began a faculty position at the University of Chicago where she founded the Institute for Mind and Biology.

In 1998 McClintock and her colleague Kathleen Stern reported evidence for a direct influence of human sweat on menstrual cycle timing. McClintock and Stern took samples of underarm sweat—secretions thought to contain human pheromones—from a group of women when they were ovulating and then again when they were in the preovulation (follicular) phase of their cycles. These samples were applied to the upper lip of 20 other women each day for four months: two months with the ovulation sweat and two months with the preovulation sweat. Menstrual cycle timing was assessed by measuring hormone levels in urine samples taken from the women nightly; a method which avoids the problems inherent to self-reports. Results revealed that when the women were given the ovulation sweat, their cycles lengthened by two days. When, instead, the recipients were given the preovulation sweat, their cycles *shortened* by two days. This study is considered one of the more rigorous demonstrations of how sweat, and the putative pheromones it contains, can influence menstrual cycle timing.

Still, the controversy surrounding human pheromones persists. To understand why, it is important to see how pheromones have been defined for animals. Pheromones in the animal world are typically considered to be of two types. *Releaser pheromones* are the chemical signals emitted by one animal that, when detected by another animal of the same species, cause it to immediately perform a set behavior. The fear scent that induces spontaneous scattering of ants would be an example, as would the secretions in a male pig's saliva that cause a female in heat to assume a submissive copulatory position. In both cases, the chemical influence seems to induce immediate, almost reflexive, behaviors.

In contrast, *primer pheromones* are the chemical signals that induce more long-term physiological changes in the receiving animal. The secretion of female fish that stimulates sperm production in males constitutes a primer pheromone, as does the chemical that induces the aforementioned menstrual cycle synchrony in cohabitating mice. Common to both releaser and primer pheromones is that they seem to induce specific, automatic (nonvolitional) reactions, likely through influences on the hormone (endocrine) system.

Do releaser or primer pheromones exist in humans? The evidence provided by Martha McClintock and her colleagues would seem to provide some support for primer pheromones. And other studies have provided relevant support. There is evidence, for example, that women who live with or have regular consistent social contact with men have shorter and/or more regular menstrual cycles (an observation first reported in McClintock's 1971 paper). A related study has shown that regularly applying a man's sweat sample to the upper lip of a woman will also regularize her menstrual cycle. Because men's sweat contains a steroid (androstenone) known to act as a pheromone in other species, this research has been taken as support for primer pheromone influences on a woman's menstrual cycle. Still, this research has not been without criticism. Typically the criticism centers on the number of subjects tested and the types of statistics used to evaluate the results.

■ ■ ■ ■ ■

BUT WHAT OF RELEASER pheromones? Might they influence our behavior? Recall that for animals, releaser pheromones seem to induce immediate behavioral changes such as sexual responsivity postures in pigs and scattering reactions in ants. It's likely that nothing as extreme as these reactions occurs for us. But what of less dramatic reactions to releaser pheromones? Can pheromones change, for example, how often we have sex?

This would be a major coup for the fragrance industry. And there have been claims of pheromone influences on sexual behavior from both commercial and scientific sources. However, these claims have been strongly criticized on scientific grounds. To date, there is no solid evidence that applying synthetic pheromones can change sexual behavior, so don't go spending your money yet.

But consider another potential human releaser pheromone: one that can induce a fearful response. While it is unlikely that a fearful odor can make us scatter like ants, what about that startle response? Recall that undetectable sweat taken from a fearful individual can amplify our eyeblink startle response when we listen to loud, abrupt sounds. Should this fear-induced sweat odor be considered a true releaser pheromone? Probably not. While the fear odor amplified the blink response, it didn't directly *induce* the response: the loud noise served that purpose. And while an eyeblink could be considered a "defensive posture" (from projectiles), it is of a very different magnitude than the pheromone-induced scattering behaviors of other species. But still, the automaticity of the startle reflex could mean that fear-induced sweat odors provide a more subtle type of releaser pheromone for us—one that has been tempered by other evolutionary pressures.

There is one more possibility for a human chemical signal that could act as a true releaser pheromone. As you learned in the last chapter, newborn infants are able to recognize the smell of their mothers based on secretions from the glands around the nipples. This skill likely helps the infants find their primary source of nutrients, their mother's breast milk. The research suggests that the odor of a mother's nipple can induce an automatic behavior in her infant. It turns out that if a *three-day-old* infant is placed on a changing table eight inches from a small pad that had been worn on its mother's nipple, the infant will spontaneously wriggle toward the pad. In fact, nearly every infant that has been tested in these experiments makes it to the pad within two minutes: an impressive feat for three-day-olds. This spontaneous wriggling toward the sweat-infused pads could be a true releaser-pheromone induced behavior.

Of course it is difficult to compare these infant responses to the seemingly reflexive behaviors displayed by ants and pigs to releaser pheromones. We tend to think of human behaviors as more complex, more conscious, and less reflexive than those of other animals. In fact, these differences in how we typically evaluate human and animal behavior have led some researchers to a revised definition of pheromones.

You Know the Odor of Compatibility

WHEN I ASK MARTHA MCCLINTOCK why she thinks there is so much controversy surrounding human pheromone research, she offers a few speculations. "The thought of the commercial abuse of pheromones is frightening. It's difficult for people to separate that fear from the actual data. Also, I think it's intimidating to think that even a very small part of our sexual behavior and mate choices could be based on something as basic as smell. When it comes to sex and love, people really want to believe they're in control."

And McClintock along with many other human pheromone researchers believe that too much emphasis has been placed on the "animal model" of pheromones. They argue that the criteria of pheromones inducing either reflexive changes in behavior or longer-term physiology is too limiting—for both humans and animals. They believe that pheromones can have additional effects that are less observable and less reflexive. And while these more subtle effects may be more observable in humans—based simply on our ability to convey introspective experience—animals are also influenced by pheromones in these more subtle ways.

The modified definition of pheromones proffered by McClintock and others includes two other classes besides the releaser and primer types. *Signaler pheromones* (borrowed from the insect pheromone literature) are thought to provide information about the sender including its gender, kin relations, and reproductive potential. *Modulator pheromones*, on the other hand, are thought to influence mood and emotion in the receiver. Based on this expanded definition, a number of the phenomena already discussed in this and the preceding chapter could be considered pheromone-influenced. These include the ability to recognize kin through smell as well as *all* the influences of fear-induced odor discussed above.

And there are many other olfactory influences on your moods, preferences, behaviors, and brain that could be under signaler and modulator pheromone control. Recall that the major histocompatibility complex, MHC, is considered the gene set bearing on your scent signature, allowing you to be recognized and associable to kin by smell. MHC also has the critical function of providing your immune system with the necessary code to discern pathogens in our bodies. The more extensive this code, the healthier—or *immunocompetent*—you are likely to be. And the extensiveness depends on the *dissimilarity* of MHC between your mother and father. Thus, the more dissimilar the MHC code between your parents, the more extensive your own MHC code and the more immunocompetent you are likely to be. Put simply, MHC-dissimilar parents generally give birth to healthier offspring.

It would therefore behoove nature to bring MHC-dissimilar individuals together to mate. And nature does this through smell. Animals as diverse as mice, birds, fish, and lizards prefer mates with MHC dissimilar from their own, and detect this dissimilarity through scent.

How about us? Do we choose mates based, partly, on the smell of their MHC? There is evidence suggesting we do. A number of T-shirt tests have found that both men and women find the smell of opposite-sex individuals more attractive if the donor's MHC differs from their own. There is also some evidence that married couples have less similar MHC compositions than would be predicted from chance (but this has not always been replicated). Interestingly, for women, sensitivity to MHC compatibility can be severely degraded by use of an oral contraceptive. Women on oral contraception usually find the odor of men who are MHC-similar to themselves more attractive. This could account for some of the conflicting results on married couples.

There is also recent evidence suggesting that, at least if you're a woman, your sexual behaviors and fantasies may be influenced by MHC. Based on questionnaire responses and MHC analyses (from saliva) of dating couples, the similarity in MHC between the partners strongly predicted the women's responses. Women whose partners' MHC was relatively similar to their own were more likely to: (a) be less sexually responsive to their partner; (b) have more extrarelationship affairs; and (c) be more attracted to men other than their partners. Further, these tendencies were amplified during the most fertile phase of the women's cycles. The men's responses weren't influenced by MHC similarity in this study. Whether this is related to men's less sensitive olfactory skills or some other male failing is unclear.

While no direct test of olfactory preferences was conducted in this study, it is a sensible assumption that the dependence of the women's sociosexual behaviors on MHC similarity was mediated by smell. After all, MHC is known to be conveyed through odor in other animals, and predicts odor preferences in humans. And the fact that the women whose partners were MHC-similar showed greater sociosexual "distraction" during their most fertile phase also indicates an olfactory basis to the phenomenon. Because as you'll see, a woman's odor preferences change in fascinating ways during her most fertile phase.

You Know the Smell of Symmetry

WHEN WOMEN ARE IN their fertile periods, they prefer the smell of a symmetric man. Take a good look at your two wrists and compare their sizes. Hopefully, their widths will seem pretty much equivalent. You can also compare the widths of your two elbows, ankles, knees, and the corresponding fingers of your two hands. For most of us, it would be hard to detect any differences when simply eyeballing these corresponding body parts. But as it turns out, people are subtly different on the two sides of their bodies, and the extent of these differences can indicate one's genetic health. It's long been known that low body symmetry in animals can be a result of genetic abnormalities caused by in-breeding, mutations, and other factors. Low body symmetry in animals is predictive of slower growth rates, reduced longevity, and reduced fertility.

But over the last 15 years, evidence has appeared suggesting that body (and face) symmetry can predict *human* health as well. Low symmetry is predictive of worse genetic, physical, and mental health, as well as depressed cognitive skills and IQ. In contrast, research shows that men with *high* symmetry (more equal wrists and ankles) typically have more sexual partners, quicker access to romantic partners, and induce more copulatory orgasms in their mates. So, it turns out that size does matter after all: bilateral equivalence of appendage size, that is.

And things get even more interesting. Because the degree of your symmetry is an inherited trait, it not only provides a marker for your genetic health, but it is also a marker for the potential health of your offspring. It makes sense then, for members of a species to search out mates with high body symmetry. But it is often difficult to see body region symmetry with the naked eye, so nature has apparently compensated by making body symmetry something that we can also *smell*. Of course, it's impossible to smell the relative symmetry of someone's wrist, per se. But it seems that a woman in her fertile phase can smell the odor effects of the physiological processes related to a man's body symmetry.

For a woman in her fertile phase, the degree of a man's body symmetry strongly predicts how pleasant and sexy she finds the man's smell. A more symmetric man smells better to the fertile woman. A woman not in her fertile phase doesn't have this preference. It's likely that the men's odor contains a signaling component that informs women about the genetic health of potential mates and offspring. It's also likely that the biochemical effects of a woman's fertile phase induce a heightened olfactory sensitivity to this component.

A woman's fertile phase also affects her smell preferences in, shall we say, more animalistic ways. Male dominance is an important factor in how female animals choose mates. From mice to apes, the more aggressive, bullying males tend to have more sexual partners. Of course as humans, we like to think of our tastes in suitors as more sophisticated. But not if our noses had their way. Women in their fertile phase rate the odors of more dominant men as sexier than the odors of less dominant men. (This is particularly true of women in relationships.) These results could mean that a woman, at least during her fertile phase, can detect a dominance-signaling component available in male sweat odors.

Thus, it could very well be that men provide a signaler pheromone to fertile women, indicating their reproductive potential. And women seem to partially reciprocate the favor. Three recent studies have shown that men can detect (underarm) odor differences between women in their fertile and nonfertile phases, and find the former odors more pleasant. Interestingly, men also find underarm odors collected during the menstrual phase of a woman's cycle as the least pleasant and attractive of the set. Perhaps the menstrual phase odor provides a subtle cue for men to not engage in intercourse during this time, thereby reducing the health risks inherent to menstrual-phase intercourse.

Regardless, the fact that men prefer the body odor of women in their fertile phase suggests that the odor contains a signaler pheromone. The findings also challenge the long-held assumption that humans, unlike most other animals, conceal their fertile phases. While human fertility may not be advertised in ways as dramatic as the genital swelling and facial color changes of apes, or beak color changes in some birds, the new results suggest that information for fertility is available through odor. And as you'll learn in Chapter 11, new research shows that a woman's fertility may be inadvertently advertised in other ways as well.

In sum, there is now evidence for a signaling component in men's and women's odors that can influence a receiver's odor preferences and, possibly, behavior. This evidence for signaler pheromones is supported by physiological research. Chemical compounds related to putative pheromones (derived from male sweat and female urine) can induce changes in a receiver's body temperature, heart and respiration rate, and skin moisture. These same compounds can cause changes in a receiver's brain regions associated with emotion. In a particularly compelling example, a compound found in male sweat has been shown to modulate the brain mechanism known to promote ovulation in women. In all of these demonstrations, the chemical compounds are presented to subjects at subliminal levels.

The Odors of Others Can Affect Your Mood and Arousal

But what of *modulator* pheromones, those thought to induce changes in your emotion and mood? You've already learned that fearful odors can influence your interpretations of facial expression, eyeblink reactions to loud noises, and accuracy in identifying related words. But these effects may not involve mood changes per se, as a true demonstration of modulator pheromones would require.

In fact, there *is* evidence suggesting that putative human pheromones can influence your mood. In another study conducted by Martha McClintock (with her colleague Suma Jacob), a steroid compound related to male sweat was found to heighten the moods of women. The women's moods were significantly elevated despite their inability to detect the presence of the steroid.

Other research has shown that inhaling a male steroid can enhance women's emotional reactions to happy and sexually arousing film clips. (While the steroid's odor was detectable for the women, it was indistinguishable from a nonsteroid control odor that was also administered.) The male steroid also served to elevate physiological measures in the women including their heart rates, skin moisture, and *cortisol* levels. The fact that the male-steroid odor elevated the women's cortisol is significant. Cortisol is a hormone whose increase is often accompanied by elevated mood and arousal. These results show that a specific component of male sweat can actually modulate a female's endocrine system—the system responsible for hormone regulation. While prior research by McClintock had shown that low-level sweat odors can influence the hormone regulation controlling menstrual cycles, a specific chemical compound found in male sweat had not been shown to influence female hormones.

This study also provides an encouraging result for the fragrance industry: women reported greater sexual arousal when they received the male steroid odor. This is a provocative finding that has now been replicated in a number of other studies. However, the implications of these findings should be made very clear. It's likely that the odor itself did not *induce* sexual arousal; that probably came from watching the sexy film clip. Instead, the male steroid odor subtly enhanced this arousal, perhaps not unlike the effects of listening to Marvin Gaye or some other romantic music.

One final study on sexual arousal should be discussed before taking stock in the human pheromone story. Male steroids are not the only odors that can make a woman feel a little sexier. It turns out that the smells of nursing—yes, *nursing*—can do the same thing. McClintock and their colleagues collected odor samples from lactating women by having them wear gauze pads on their nipples and underarms. These samples were given to women who then applied them to their upper lip every morning and evening for three months. The women also kept a daily "sex log" in which they rated their sexual activity, as well as their sexual desires and fantasies. The results showed that relative to a control group of women, women receiving the lactation sweat odors reported increases in sexual fantasies and desires. They did not, however, report an increase in sexual activity itself—a fact that will be discussed soon.

Why would the smells of nursing, of all things, heighten sexual feelings in women? The likely answer is a reminder of our animal origins. In the animal world, it's generally thought that coordinating pregnancy and lactation across neighboring females raises the chances of offspring survival. Groups of pregnant and nursing animals can share resources and protection. While coordinated pregnancy and nursing in animals is dependent on seasonal resources, there is also evidence that pheromones emitted from one female can change menstrual cycle timing in another. But what about us? Despite the complexities of a modern world that may no longer benefit from coordinated pregnancy, there is evidence (albeit controversial) that the smell of a lactating woman can influence menstrual timing in another. If true, then it is not hard to imagine that lactation odors may also induce an increase of sexual receptiveness in other females, for similar reasons. Our animal heritage of coordinated pregnancy and rearing could then account for the subtle sexual influences of lactation odors reported by McClintock and her colleagues.

THE INFLUENCES OF SUBLIMINAL ODORS ARE SUBTLE AND COMPLICATED

SO WHERE DOES THIS LEAVE US? Do human pheromones exist? The answer depends greatly on your definition of pheromones. If your definition is restricted to chemical signals that induce releaser- and primer-type influences, then the evidence for human pheromones is incomplete. There is evidence for primer-type influences on women's menstrual cycles, but this evidence is still somewhat controversial (albeit less so as research progresses). And evidence for true releaser pheromones in humans is scant, with the one possible example being the "reflexive" crawling of neonates toward their mother's nipple scents.

But if your definition of pheromones is broadened to also include the signaler and modulator types, then their existence seems more likely. There is mounting evidence that human odors can *signal* an individual's fearful state, immunocompatibility, kin relations, body asymmetry, male dominance, and menstrual phase. There's also evidence for physiological and neurophysiological responses to many of the signals. And there is research evidence for human modulator pheromones. Steroids associated with putative male and female pheromones have been shown to modulate mood and sexual thoughts, especially in women. There is also evidence for physiological effects that accompany these mood changes.

That said, should the signaler and modulator influences be considered true pheromone effects? It's unlikely the debate will be settled soon. The debate hinges on deeper issues concerning thought, emotion, and free will in humans versus animals. In this regard, Martha McClintock provides an interesting insight: "We have research showing that the effects of androstadienone [a putative pheromone] influences the moods of women more or less depending on who else is in the room. Context is absolutely critical. But does this mean that it's not a pheromone? Well, if you go back and carefully read the research on pigs, it turns out that the female's mating stance is dependent on more than just the presence of the pheromone in the male's saliva. It also depends on the female *hearing* the male nearby—and even then, it doesn't work every time. So for the pig, and for us, context is critical. Our view of pheromones as producing reflexive behavior in animals is probably too simplistic."

So, scent influences between animals might not be that different from those between people. In both cases, scent effects are just part of the equation (albeit a greater part for animals than for people). And this is an important point to remember when considering all of the subliminal effects discussed in this chapter. While your unconscious ability to integrate odors is astounding, it should not be considered determinate. Subliminal odors cannot cause you to buy things you don't want or choose mates you don't like. Certainly every other sense, including the "common" type, plays a critical and ultimate role in your decisions.

Part III

Tasting

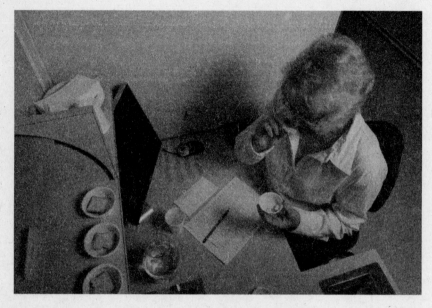

Professional food tasters can refine their palates to predict when a product's cooking oil will turn.

CHAPTER 5

Cold Leftovers with a Fine North Dakota Cabernet

"ARE YOU FEELING BETTER NOW?" THE HOSTESS ASKS.

"I'll feel better once I finish this," the young woman smiles nervously as she holds up her glass of wine.

The hostess then comes over to us and whispers, "This happens about once a night. Most people go back in, but once every few weeks someone's too scared to return."

This is not what one wants to hear while waiting to enter a restaurant.

But this is no ordinary restaurant. My companions and I are in the lobby outside a dining room that is pitch black. We have been told that once we enter, we will be unable to see anything, including our server, each other, and our food. We should expect to feel disoriented and perhaps a little frightened at first. This explains why the young woman has stepped back into the dimly lit lobby to finish her wine. But, we're told, most customers start feeling comfortable after a few minutes, and ultimately feel more relaxed than they would in a normally lit restaurant. The big question is whether our experience of the food itself will be enhanced, without the distractions of vision.

Before entering the dark dining room, we are asked to order our food from a specialized menu. The menu is configured so that we can choose how much information we get about what we'll eat. We can choose to see, and select from, the full detailed menu, as we would in most restaurants. Alternatively, we can choose to know nothing except for our selection of a meat, fish, poultry, or vegetable entrée. Or, for the truly adventurous, we can decide to be told nothing whatsoever about what we will be served. My companions look at the still-panicked expression on the woman nursing her wine and opt to know exactly what they'll be eating. I try my luck, and ask to be served the vegetable entrée, whatever it is.

We are greeted by our server, who, like most servers in the restaurant, is visually impaired. It is thought that their experience relying on their other senses allows them to more easily navigate the hazards of a totally dark, very cluttered restaurant. Our server instructs us to place a hand on her and each others' shoulders so that we can be led through the restaurant to our table. This involves going through double doors and then a series of very sharp turns, all methods to ensure that no light gets into the main seating area. These methods are totally effective. As we finally arrive at our table, I realize that I can see nothing: everywhere I look is totally black. It is the darkest place I have ever been and it feels odd to notice no visual difference with my eyes opened or closed.

Our server helps us acclimate to the table by guiding our hands to our silverware and glasses. As she leaves to get our food, my companions and I chat, and become more comfortable with the dark. We start to notice some interesting perceptual effects. We reach out our hands to get a sense of how close we are to each other. From this, we realize that we've been experiencing ourselves as seated much further apart than we actually are. Also, the background conversations we hear from other tables are noticeably garbled. Despite our attempts to actively eavesdrop, it's virtually impossible to recognize what's being said. It sounds like the indistinct "warble" produced intentionally by extras on a movie set to simulate the background sounds at a restaurant or party. This effect is likely a result of not being able to *look* at the conversants in a way that normally facilitates eavesdropping (see Chapter 10). Becoming aware of these strong perceptual effects, we wonder how the dark will also affect our sense of taste. And then our food arrives.

■ ■ ■ ■ ■

WHY DOES FOOD HAVE FLAVOR? Food should just afford passage into our digestive systems, or be chewable to allow this passage. This process would simply require our touch sensations (broadly defined) allowing control over food using our lips, tongue, and teeth. In this sense, the mechanics of eating and swallowing are similar to our hands squeezing a tube of toothpaste. Our sense of touch is all the sensory information we need to control squeezing and swallowing.

But our senses get so much more from food. The myriad flavors we experience from food seem endless. And our passionate need for these sensory experiences can lead many of us to weight and health problems; problems severe enough to be deadly. Flavor seems important for food, and there is a reason for this.

Food has flavor so we know not just how to eat but what to eat. Research in anthropology and evolutionary biology shows that our ancestors developed sensitivities to flavors so that they would seek out nutritional foods and shun the toxic. For example, our tongue's sensitivity to the primary tastes of sweet, salty, bitter, sour, and umami (savory) is likely a result of nutritional foods providing some of these flavors, and toxic foods providing others. We likely crave sweets due to the nutritional benefits provided by fruits, and dislike strong bitter tastes because of the toxins in poisonous mushrooms and other bitter-tasting plants. Similar adaptive causes have been used to explain our reactions to the other primary tastes.

Of course, our modern processed foods exploit these flavor preferences. This leads us to often eat deceitful foods, which to our sense of taste promise much more nutritional benefit than they deliver. And when our access to food exceeds our bodies' physiological cost to finding food, weight and health problems ensue. But don't blame your sense of taste. That sense evolved to work with a body that needed to find natural food, at a significant metabolic cost.

The adaptive importance of choosing ingestibles that provide nutritional benefit, and little toxicity, has shaped a flavor mechanism that recruits all of the senses. While you may experience taste as something your tongue does, your nose, eyes, ears, and skin are all involved. Any master chef will tell you that flavor is dependent on how food smells, feels, looks, and even sounds. This provides the chef with a culinary palette to create endless combinations and permutations for your tasting pleasure. And as the research on taste shows, cross-sensory influences can create emergent flavor experiences that are different from the sum of their parts—and are sometimes illusory.

■　■　■　■

IF YOU'VE NEVER NOTICED the variety of shapes that lettuce can take on, try eating in a pitch-black restaurant. It's one of the first things my companions and I notice about eating in the dark: we are much more aware of the shapes of our food—shapes we are apprehending with our mouths. This is an interesting but not altogether pleasant experience. The bread, on the other hand, is warm and comforting, even if it does feel a little spongier than we would normally notice. There is no question that the dark has made us more aware of how we feel our food.

And then our entrées arrive. My dish smells a little exotic and spicy, but not unfamiliar. I tentatively place my fork on the edge of the plate and scoop a small amount of what feels to be a thick, pliable substance. Once it's in my mouth, I establish that it's rice, probably brown—based on its texture—and covered with a Thai-style curry. I decide to stab my fork somewhere in the middle of my plate. I spear something and bring it toward my mouth. But long before it reaches my lips, something sharp sticks the outside of my cheek. I pull it back, reorient my fork and place it in my mouth. It's a full-sized green bean, very lightly steamed and flavored with the same curry sauce as the rice. As I bite down, the sound of its crunching seems especially loud, highlighting its brittle texture. It's an interesting experience, but it seems to overwhelm the actual flavor.

Later, my companions and I realize that we've all had a similar experience with our food. We agree that everything tasted a bit bland. While this could be a failing of the chef, it seems odd that everything we've tasted, including the bread and our different appetizers, entrées, and desserts—13 different dishes in all—lacked flavor. We also agree on how unlikely this would be for an upscale, albeit unique, restaurant. While it might not be surprising for some dishes to be weak, the fact that all of the dishes seemed to lack flavor strikes us as odd.

But this is something we could have anticipated from the research on taste.

YOU TASTE WHAT YOU SEE

RESEARCH SHOWS THAT WHAT you see, or don't see, strongly influences what you taste. Certainly, you are aware of how the appearance of food can influence how appetizing you find it. But you may not know that a food's appearance strongly influences how you experience its flavor. For example, if you were presented with four glasses of commercially available fruit drinks (cherry, orange, lime, and grape), you would likely be able to identify all of the flavors correctly all of the time. If, however, you were unable to see the colors of the different drinks, your ability to identify the flavors would drop to as low as 20 percent correct. This would be true even when nothing had been done to the drinks' smells and tastes. Further, if the colors of the drinks were switched so that the cherry drink were colored orange, the orange drink were colored green, etc., your perception of the flavors would be affected by the drinks' colors. A lime drink colored orange would more likely taste orange than lime, and an orange drink tinted green would more likely taste lime than orange.

And, most relevant to my own experience in the dark restaurant, your ratings of the flavor *intensity* of the drinks would also be strongly influenced by their appearance. For example, if a weak drink were given a bright color and a strong drink, a dull color, the color intensity would largely determine how strong the drinks *tasted* to you. The influence of color can even induce a flavor that is not there. When water is tinted orange, most people experience its *flavor* as orange. Conversely, if the strongest orange-flavored drink is given no color (is clear), it almost always tastes as if it has little if any flavor. Thus, your sense of taste can be strongly influenced by what you see. And if you don't see enough of what you're used to, flavors can be severely muted. This could very well be the reason my companions and I found the food in the darkened restaurant consistently bland.

Lest you think these results are specific to artificial fruit drinks, similar results have been found with cakes, cookies, sherbet, custard, and candy. And color is not the only visible characteristic that can influence what you taste. Visible texture can influence how creamy and flavorful you rate a custard. Appearance can also influence your ability to simply *detect* that a flavor is present or that two flavors are different. Finally, the influences of vision on flavor seem to be truly *perceptual*, and not something that outside knowledge can undo. It turns out that subjects report taste changes even if they are told that the taste and color are discrepant. You can't help the fact that your eyes "taste" food.

All of these results are consistent with neurophysiological research. Research on primates shows that the same cell groups that respond to taste also respond to *seeing* food. Related research shows that seeing food can induce activity in the hypothalamus, a region known to control appetite and eating-related behaviors. Brain scanning studies on humans show that when smelling a food to judge its edibility, visual centers are automatically recruited even if no visual stimulus is present. This neurophysiological evidence attests to the importance of visual input to your perception of taste.

You Taste What You Hear

It may seem surprising that vision can have such strong influences on taste, but not as surprising as the fact that *sound* can influence what you taste. Research shows that the sounds we hear when we chew influence how fresh and pleasant a food tastes. For a variety of foods including fruits, vegetables, and crackers, the degree to which chewing sounds can be described as "crisp" is closely related with how pleasant these foods taste. In general, we like crisp foods and largely experience crispness as a dimension of *taste*. Yet, research shows that we rely heavily on sound to determine crispness. In fact, ratings of food crispness are the same whether we are chewing the food ourselves or simply listening to someone else chew.

So, how does sound affect what you taste? As you eat, different foods make different sounds. These sounds reach your inner ears through two routes. First, there is the common way, via air disturbances that travel from your mouth out into the surrounding air and then around to your ears. Second, there is *bone conduction:* mechanical vibrations conducted through your teeth, jaw, mandible, and skull directly to your inner ear. The sound traveling through both paths can influence how you hear your food, with the relative importance changing as your lips open and close to chew. But the outside sound entering through your outer ears is easier to experimentally manipulate, and that is what researchers have done to understand how sound affects taste.

In an experiment that won the coveted Ig Nobel Prize, subjects were asked to bite on 90 potato chips that, unbeknownst to them, were all of the same size, shape, thickness, and texture (Pringles: The Scientist's Chip of Choice). Subjects bit each chip once, and then spit out the severed pieces. A microphone was placed in front of the subjects' mouths so that the biting sounds could be modified and played back live to subjects via headphones. Sounds were modified electronically by independently manipulating the loudness and timbre (brightness) heard by subjects. Subjects were asked to judge how crisp and fresh each chip tasted. They were told nothing about the sound manipulations, the uniformity of the chips, or that they should base their judgments on any particular sensory modalities.

Results showed that the biting sound subjects heard had a significant effect on how the chips were rated. In general, the louder and the brighter the sound, the fresher and crisper the chip tasted. If the sound was quieter and duller, the chip tasted stale and soft to subjects. Recall that there was no manipulation of the chips themselves (Pringles are Pringles). Only the sound was changed, and that was sufficient to influence how fresh and crisp or stale and soft the chips tasted. In fact, at the end of the experiment, a majority of subjects thought that the chips were selected from a number of different packages and brands. They were surprised to learn that the chips were all the same. Like vision, hearing can determine what we taste and what we think we've put in our mouths.

You Taste What You Touch

Linda Blade has trouble with leftovers. Not all leftovers, just those containing meat of any type. She refuses to reheat a meat dish and will choose instead to eat cold Chinese leftovers on warm rice, or cold meatballs on warm pasta. The reason is that she can't help but taste what she calls the "warmed-over" flavor of reheated meat. As a *sensory specialist* working in the field for over 20 years, she has been trained, and trains others, to detect the undesirable flavor of warmed-over meat—a flavor that food manufacturers try to avoid. The manufacturers pay Blade well for evaluating, and training panels to evaluate, the flavor of their foods. The panelists work to determine a flavor's components, consistency, and quality. Fortunately, when Blade eats outside the lab, warmed-over flavor is one of the few quality problems she still can't ignore.

But this wasn't always the case. For the first few years after her own taste training, she had trouble turning off her analytical palate. She remembers difficulty ignoring the flavors of burnt peanuts in peanut butter and of early oxidation of the oil used to cook potato chips. This is typical of newly trained sensory specialists. "The first few years after you train, casual eating is not a pleasant experience. And if you ever end up going to a restaurant with a bunch of new sensory evaluators, God help you," she laughs. Much of the evaluators' training is geared toward picking out diminishing quality before it can be tasted by the general public. They are the ones largely responsible for determining the projected shelf-life of a food—the "Best eaten before . . ." dates printed on packages. "We'll need to detect the slightest hints of oil starting to turn bad in snack foods." This takes intensive training to taste, and this training is difficult to turn off at first.

But despite this temporary drawback, tasting panelists typically enjoy what they do. According to Linda Blade, "They love learning about their own taste skills. They're amazed at how complex their experiences of flavor can be." When Blade is hired by a manufacturer (Kraft, Cadbury, Pepsi), she recruits panelists from the local community. No special skills are required, but applicants are screened for a simple ability to recognize the flavors of sweet, salty, bitter, and sour, and to taste different concentrations of these flavors. After passing this screening, their taste training begins.

"Basically, I'm training them to become human tasting instruments. They're learning how to concentrate on individual flavors, and just as importantly, acquiring a language to describe their flavor experiences," Blade says. "Most people don't think very much about taste. They eat something and they either like it or they don't. But taste is much more complex than this. My job as a trainer is to get people to *really* taste things, and then to report what they taste in a consistent way." Much of the training involves having panelists taste and sniff ingredients—cinnamon, clove, melon—and try to recognize them. Tastes, like smells, are notoriously hard to recognize without some context (visual, tactile), and novice panelists are often faced with tip-of-tongue experiences. "They'll say that the taste is familiar, but not be able to identify it with words." Blade helps out by encouraging them to think of some personal association. "A flavor might remind someone of a cake their grandmother used to make. I'll then tell them it's clove, and that whenever a taste reminds them of their grandmother's cake, they'll know they're tasting clove." From weeks of this type of training, most panelists will be able to identify and quantify flavors in a food, as well as evaluate the flavors' freshness and consistency across multiple samples. And panelists turn out to be impressively consistent with these evaluations, both across their own judgments and those of other panelists.

When I ask Blade what it takes to become an expert taster, she responds that it really only depends on a panelist's motivation and ability to focus. She believes that there is nothing inherently different in the tasting physiology of the expert and that, in principle, we all have what it takes to be expert tasters. So, it simply comes down to attention and practice, and, as you'll soon learn, your brain's ability to support the benefits of intensive sensory experience.

Often, the sensory specialist's skills are used to evaluate a product's consistency. The ingredients of a product change whenever a manufacturer's suppliers change. Incorporating a new crop of figs into Fig Newtons changes the cookies' flavor. And for packaged products, consumers expect to taste what they're used to tasting. To ensure consistency, sensory specialists are brought in to determine in what ways a product's flavor has changed so that compensatory measures can be taken. Establishing flavor consistency in a product highlights some of the biggest challenges in sensory evaluation: methodological consistency and bias. As Linda Blade puts it, "There are *so* many things that can influence how things taste, even to the expert. We need to make sure that the size, shape, and temperature of food samples are consistent." And Blade's experience with these biases is consistent with research showing how the tactile experience of food influences what you taste.

■　■　■　■　■

THE INFLUENCES OF TOUCH on taste are extensive. Try this demonstration to determine the location of your "sweet" taste buds. Dip a Q-Tip
in a solution of strong sugar-water (mix one teaspoon of table sugar into
one tablespoon of water). Using the Q-Tip, apply the sugar solution to
the right-hand side of your tongue, starting from about halfway back.
Slide the Q-Tip along the edge toward your tongue tip while noting your
tongue's experience of sweetness. Once you've reached the tongue tip,
continue along the left side, moving toward the back. If you're like most
people, you experienced relatively little sweetness on the right-hand
side, and then greater sweetness on the tongue tip and left-hand side.
So it would seem that your sweet taste buds are on the tip and left side
of your tongue.

To confirm this, rinse your mouth with water, and redip your Q-Tip in
the sugar solution. Start on the *left*-hand side of your tongue this time.
Start from halfway back and move toward the tip, again noting the sweetness you experience. Once you get to the tip, continue along the right-
hand side of your tongue. What happened? Did you confirm that you
tasted more sweet on the left side of your tongue than on the right?
Probably not. It's likely that you tasted more sweetness on the *right-hand
side* of your tongue this time. Did your taste buds move? No—you just
experienced an illusion demonstrating how strongly your sense of taste is
influenced by your sense of touch.

In fact, you have very few taste buds of *any kind* along the left and
right sides of your tongue. You do, however, have a high concentration of
taste buds on your tongue tip, and activating them with the sugar solution caused your awareness of the flavor's location to then follow where
your tongue was being *touched*, instead of where your sweet receptors
are located.

This *tactile capture* of flavor is actually very common, likely occurring
whenever you eat. It occurs when you experience a hard candy's sweetness as following the candy as it moves around your mouth. A majority
of the locations the candy occupies do not have sweet taste buds. Still,
the candy seems to taste sweet wherever it lands in your mouth. Tactile
capture also explains an interesting phenomenon occurring for individuals who have lost taste bud function on one side of their mouths (usually

due to illness or accident). Most of these individuals have little, if any, awareness of their deficiency, likely because flavor seems to follow food wherever it touches the mouth.

Why does tactile capture occur? It's likely a result of your perceptual systems' general strategy to perceive things as unitary and coherent (see Chapter 11). The fact that you have taste buds spread inconsistently throughout your mouth could, in principle, render a single piece of food as either elusive or inconsistently flavored. Tactile capture creates a generalization of flavor location to help you perceive the food piece as a single object that carries flavor. This helps allow you to experience food—and not your tongue—as having flavor.

Of course you use your sense of touch to determine more than where food is located in your mouth. You use it to determine a food's texture, temperature, and irritants if it's spicy (although temperature and irritation are formally based on chemical effects). And as Linda Blade has found with her expert tasting panels, each of these characteristics of touch influences what you taste. You are certainly aware of how a food's texture bears on how appetizing you find it. No one likes pulpy apples or lumpy milk. But your sense of a food's texture or viscosity also influences how that food actually *tastes* to you. More viscous, thicker textures cause a food to have weaker taste than less viscous textures. This has been shown in research on gelatins, custards, tomato and orange juice, and even coffee. Critically, the influences of food thickness on flavor are not just based on how viscosity physically affects the speed with which chemicals are released from the food (thicker foods release odorous and flavorful molecules more slowly). Even when controlling for these factors, food thickness has a strong perceptual influence on flavor and this is something food manufacturers take into account in their ingredients.

Temperature also influences how food tastes. You've no doubt experienced the unpleasantness of tasting a food that is served either too cold or too warm. Restaurant dinners are sent back more often for being too cold than for any other reason. But beyond the general sensual unpleasantness of tasting a cold penne alfredo, or a warm soft-drink, your sense of taste *itself* is influenced by a food's temperature. Research shows that your sensitivity to taste is at its peak for foods served at a temperature close to that of your tongue (which ranges between 71.6 and 98.6 degrees Fahrenheit). Fine restaurants know this and often test the temperature of their dishes to determine the optimal delay between when a food leaves the flame and arrives at your table. For the expert tasters, deviations of a couple of degrees can influence how something tastes.

But temperature can influence taste in more mysterious ways. Try this demonstration. First, prepare a small glass of warm water. Now take an ice cube from your freezer and place it on the underside of your tongue near the front or sides. Do you notice any flavor? If you're like most people, you'll experience a subtle and temporary salty or sour taste. Now move the ice cube around to the other parts of your tongue. You might also experience other subtle flavors as you do. But, now with your tongue nice and chilled, take a sip of the warm water and hold the water on your tongue. Any flavors? Most people experience a subtle hint of sweetness when their tongues are quickly warmed up like this.

Why do these effects occur? Ice cubes and warm water have no intrinsic flavor, so the effects are a result of the temperature changes on your tongue's surface. It turns out that temperature itself can change properties of your taste bud receptor cells and induce cell activity that your brain interprets as taste. The effects show how food temperature can directly influence your sense of taste, even when that food has no intrinsic flavor. This is consistent with neurophysiological evidence for the existence of temperature-sensitive neurons in your brain regions associated with taste perception, as well as in the pathway from your tongue to these regions.

Finally, your sense of taste depends on how much pain you're in. The touch sense of your internal mouth is affected by the irritants found in foods containing spicy peppers or mustards (wasabi). These irritants affect the sensitive touch fibers on the mucous membranes of your inner mouth—or, for wasabi, the trigeminal nerve in the back of your throat and nasal passages, also considered a *touch* receptor. Yet, the irritants don't act to induce pain in the taste buds themselves. In this sense, spicy foods don't affect your taste receptors, but they do affect your touch receptors. What you touch is what you taste, even if what you touch is painful.

Of course for many of us, a little pain can be a good thing—possibly releasing those ever-popular endorphins. This can make our eating experience more exciting and seemingly heighten our senses. But initially placing spicy food in your mouth does just the opposite to your sensitivity for flavors. Research shows that one of the most common compounds in spicy food, capsaicin, reduces your sensitivity to sweet, salty, sour, and bitter flavors. (New research is showing that capsaicin could be used as an effective topical anesthesia.) Like all pain, though, the best part is when it stops, and when the effects of capsaicin wane, there is evidence that your sensitivity to certain flavors can actually be *enhanced*. While this evidence is preliminary, it could help explain why fans of spicy food feel that their pain actually enhances food flavor.

Of course, too much of a painful thing can produce discomfort and numbness. Fortunately, there are salves. Sucrose can reduce the burning sensations produced by capsaicin, so it is wise to drink a Thai iced tea, or sweet lassi or horchata or soft drink with your spicy food. Dairy products also work well, and alcohol has been found to help a bit. But while you might think cold water would put out the fire, it has been found barely effective at reducing capsaicin burn.

You Taste What You Smell

NOTHING INFLUENCES TASTE AS MUCH as smell. Much of what you experience in your mouth is actually in your nose. Try this stinky demonstration. Peel and halve an onion and an apple so that the halves are about equal size. Now close your eyes, pinch your nose shut, and take a bite of each. You probably won't be able to taste the difference. Now try the same demonstration without plugging your nose. You should have *no* trouble tasting the difference. (This demonstration also works great with different flavors of gourmet-style jelly bean.)

This is essentially one of the first exercises Linda Blade has her new panelists perform. "Most of them have no idea that a majority of their flavor experience comes from smell. I tell them their tongues are only getting sweet, salty, sour, and bitter, and that everything else—lemon or chocolate flavor—is coming from their nose. After they plug their noses and experience this for themselves, they'll make the connection that food tastes bland when they have a bad cold." This nose-plugging technique is then used throughout the panelists' training. It allows them to taste the simple component flavors as registered by their tongues, and then experience how the aromatics add so much more.

Smell mostly affects taste not from your inhaling odors into the nostrils, but from the odors that enter your nasal passages through the back of your mouth. This type of smell is called *retronasal* (back of the nose) and occurs often, but not exclusively, when you're exhaling. When you exhale or even swallow, air moves from your mouth upward through your nose, and food's odors come along for the ride. The odors pass across your nasal receptors to strongly influence your sense of taste. Notice that you don't experience retronasal smelling as *smelling* per se—a fact you demonstrated when the onion's *taste* returned after you released your nostrils (allowing the air to cross your nasal receptors). You experience retronasal smell as tasted flavor, perhaps another example of "tactile capture." This is why many of us feel that there is something wrong with our tongues when we have a cold. And it also accounts for why many anosmic individuals often first seek medical help when they notice a loss of food flavor which they mistakenly attribute to their sense of taste.

The influences of smell on taste have been studied for some time, and this research has revealed some fascinating effects. For example, if you were asked to chew a flavorless gum and simultaneously have a subtle chocolate odor pumped into your mouth via a small tube, you would be convinced you were chewing a chocolate gum with a strong flavor. The intensity of a presented odor can also influence how strong a flavor tastes, and vice versa. And, just as a sweet flavor can suppress the strength of a sour taste, a sweet *smell* can do the same thing. So if you find the taste of your grapefruit juice too sour, try sniffing a strawberry as you sip—your tongue will appreciate the difference. (While it is retronasal odor that mostly influences your sense of taste, inhaling odorants through your nostrils is also effective, and much easier to manipulate.) This ability of odor to sweeten what you taste has long been used by the food industry. In fact, when you see an ingredient listed as a "sweetness enhancer," it is likely an aromatic, rather than flavorful, compound.

The integration of smell with taste is so complete that, by some estimates, nearly 80 percent of a food's flavor is determined by its retronasal odor. This is consistent with neurophysiological research showing that odor and flavor inputs converge on brain regions related to your experience of taste. And the wiring of your taste regions also allows for flavors and odors to build taste experiences that are more than the sum of their parts. Brain scanning shows that when a consistent flavor and odor are presented together, the amount of activity in taste regions is greater than the sum of the activity from the flavor and odor alone. This could explain some fascinating perceptual effects.

It's long been known that two flavors, which on their own are too weak to taste, can be presented together to produce a quite noticeable flavor. Expert chefs often make use of this phenomenon, which can actually work with up to 24 compounds too weak to taste on their own. It turns out that the same is true of smell and taste. If you were presented with a very low-level cherry aroma or low-concentration sugar solution you would be unable to detect either. But if you were presented these weak compounds together, you'd experience a sugar solution that tastes strongly like cherry. Interestingly, this effect only works when the odor and taste are consistent with one another. If the same amount of cherry odor were presented together with a weak savory-flavored solution, both of the components would remain undetectable. This suggests that familiarity with a taste-smell combination might be important for emergent flavors to occur from weak components.

The influences of smell are so profound that simply *imagining* a smell can affect what you taste. If you were asked to taste two solutions to determine which contained a very small amount of sugar, you'd be more accurate if you simultaneously imagined smelling a strawberry odor. Interestingly, if you imagined smelling *ham* instead of strawberry, you'd find it more difficult to determine which solution contained the sugar than if you did no imagining. The effects of odor on your taste are so ingrained that imagining odors affects your ability to even detect a flavor.

So the influences of seeing, hearing, touching, and smelling on your perception of taste render it the most multisensory of all your senses. And it turns out that not only do all of your senses have direct influences on taste, they all have *indirect* influences on taste in the way they influence *each other*. The smell of a custard can influence its perceived thickness, and vice versa; the texture of a yogurt changes its perceived temperature; and as you learned in Chapter 3, the color of a wine can determine whether it smells like a white or red. The results of all these indirect influences can modify how these foods *taste*.

You Taste What You Expect

ONE OF THE MOST dramatic ways your sense of taste can be changed from influences outside your tongue is from your expectations. Imagine being in this experiment. You sit down at a table and are served what looks to be a nice scoop of pink/peach-colored ice cream. In fact, the label "ice cream" has been placed on the dish to dispel any doubt. You put a small spoonful in your mouth to determine the flavor and realize that it tastes like . . . salmon. Smoked salmon to be exact because that, in fact, is its chief ingredient. Asked to describe the components of this flavor along with their intensity, subjects find the ice cream very salty, bitter, and its overall flavor strong and unpleasant. These ratings contrast strongly with those of a separate group of subjects who are served the same salmon ice cream, but are primed to taste a "frozen savory mousse" by seeing that label on the dish. These subjects find the salmon ice cream much less salty and bitter, and find its overall flavor less strong and more pleasant. So not only can your expectations influence how much you enjoy a food, but they can also influence how you experience the actual components of a food's flavor.

This is something Linda Blade has observed in her work. Before she was an independent consultant for food manufacturers, Linda worked as a food tester for a consumer magazine. There, she convened panels to evaluate the flavors and quality of name-brand products ("Which tastes fresher, Nabisco's Fig Newtons, or the cheaper supermarket brand fig bars?"). In this context, it was critical that panelists be completely naïve to the brand of food they evaluated. "If panelists recognized the brand they were eating, it would influence their experience of quality and flavor. We needed to make sure that samples were disguised and made consistent." And if this is true for expert tasters who've been trained to be objective measuring instruments, you can imagine how expected quality can influence the rest of us.

This has certainly been demonstrated in the taste research. For example, if you were poured a glass of wine from a bottle adorned with a label from a fictitious California vineyard, you would rate its taste higher than if the same wine were poured from a bottle from a fictitious *North Dakota* vineyard—a state not known for its quality wine.

More surprising would be the fact that you would rate the cheese you were given with the wine as tastier if you thought the wine was from California. It seems that what you expect about the quality of a food or beverage not only influences how tasty you find it but also how tasty you find its *accompanying* food.

This effect works in an actual restaurant environment. Customers were given a complementary glass of cabernet to accompany their prix-fixe meal. Half of the customers were told that their wine was from California, and the other half were told it was from North Dakota. After the customers finished their meals and left, the remaining food on their plates was weighed. The results were very clear: customers who thought their wine was from California ate significantly more food than customers who thought their wine was from North Dakota.

But you're forgiven for allowing your expectations to sway your enjoyment of wine and food. It's not your fault; it's your brain's. Brain imaging research shows that as subjects sip what they believe is an expensive wine, brain areas associated with pleasure are more greatly activated than if they sip the same wine but are told it's cheap. While subjective pleasure is admittedly a difficult thing to measure with brain imaging, a small brain region located behind your eyes (the medial orbitofrontal cortex) has long been thought to encode for experienced pleasantness. This region shows greater activity when a subject reports increased pleasure, whether the pleasure is induced by pleasing odors, foods, music; being touched; or having thirst quenched. And believing that you are sipping an expensive wine can increase activity in this "pleasure region" despite the wine's true quality. As far as your brain's pleasure region is concerned, you actually *do* get what you (think you) pay for.

That is, unless you're a master sommelier like Steven Poe.

You Taste What You Understand

STEVEN POE REMEMBERS HIS master sommelier blind tasting exams well. "There's a lot of pressure. You have 25 minutes to describe and recognize six wines which can be from any wine region in the world.

Essentially for each wine, you have four minutes and ten seconds to recognize its grape variety, region of origin, and vintage—while verbalizing your exact rationale for your descriptions. And there are these three masters sitting in front of you, writing down everything you say. You're totally under the gun."

The blind tasting is just one of the three phases of the master sommelier exam, the other two covering wine theory and serving decorum. And to be eligible to even *sit* for the exam, a sommelier must have gone through the other three levels of training, each with its own exam. It typically takes many years to complete all stages of training to qualify for the master's exam. Even then, taking the exam is by invitation only, and the exam itself has a 97 percent failure rate. It's not surprising, then, that there are currently only 167 master sommeliers in the world. But these masters are considered the most knowledgeable wine experts and make a very good living from both the restaurants (country clubs, distribution firms) that hire them and from training more inexperienced sommeliers. Unsurprisingly, competition can be severe, and many budding sommeliers do get frustrated by the process.

But Poe doesn't feel that the Court of Master Sommeliers (the organization that provides the training courses and oversees the exams) is an intentionally excluding group. Like most current masters, he did need to retake his master exam a few times before passing. But on each occasion, he found the adjudicating masters fully encouraging. "These are really terrific guys. They genuinely care and want to see people succeed. They do not enjoy telling people they failed."

But that doesn't mean the presiding masters provide clues or feedback while the exam is being administered. And the blind tasting is generally considered the most stressful part of the exam. It does have its rewarding moments, however, as Poe recalls. "I remember that I was working to identify the last of my six wines. Time was running out. I had looked at the wine's legs, inhaled its bouquet, and taken a sip. I was trying to describe it as if it were an overripe cabernet-syrah blend—probably from Australia. But something just didn't seem right. I thought again about its characteristics: it was a little bit off-dry, it was big, it had high alcohol, it had spiciness, and it tasted a little of dried raisins. And then it just clicked! I thought, oh my gosh, this is Amarone! [a red wine from

Veneto, Italy]. It was one of those situations where the studying and the focus just took over." Poe's insight is consistent with the research on wine experts: studying, and the explicit knowledge it brings, helps one become a better taster.

You MIGHT THINK THAT sommeliers develop more sensitive palates, allowing them to easily detect a wine's distinctive ingredients. But research shows that wine experts do not have a lower threshold for tasting a wine's components such as tannins and alcohol. Furthermore, novices seem to be nearly as good as experts at judging whether two wines are the same or different. And an expert's discriminatory skills are also limited when it comes to smelling wine. Recall from Chapter 3 that adding red food coloring to a white wine can make experts believe they are savoring the bouquet of a fine Cabernet Sauvignon.

So wine experts might not have special tongues or noses, but they may have special brains. Recent brain scans of sommeliers reveal that as they taste wine, their brains react differently from those of novices. As the sommelier sips, his brain first shows enhanced activity in the regions where taste and smell inputs converge. This enhancement likely provides the sommelier with a more vivid representation of flavor. From there, the sommelier's brain shows greater left-hemisphere activity than do the brains of novices. The brain's left hemisphere is associated with analytic processes, so its enhanced activity could provide the expert with a more intellectual tasting experience. Finally, the sip induces the sommelier's brain to show greater activity in regions associated with the higher cognitive functions of memory, language, and decision making. Increased activity in these regions can further enhance the expert's analytic tasting experience.

These recent brain scan findings are fully consistent with what is known about the *true* specialized skills of sommeliers. Wine experts have developed a rich conceptual understanding and language to describe their tastes and smells of wine. Wine affords a particularly rich descriptive terminology in providing over 500 organic compounds that bear on taste, scent, and tactile mouth feel. Extensive practice in describing the sensory effects of these compounds ("a brave little vintage with a strong acacia and peony front, leaving mirabelle plum and tobacco overtones") can lead to expertise in categorizing the different production dimensions of wine.

In turn, having explicit knowledge of grape types and production techniques allows the expert to attend to multiple flavor dimensions that are typically associated with these variables. For example, an expert is likely to be familiar with the flavor outcomes of a wine having gone through malolactic fermentation, a particular process of secondary fermentation. In a blind tasting, the expert is likely to notice one of the flavors associated with malolactic fermentation—a buttery texture, for example—and then attend to the other likely flavor results of the process including hints of yogurt and sauerkraut. This allows the expert to more easily recognize wines in blind tastings, or in the context of experiments. Thus, experts almost always show superior skills in recognizing wines they had earlier tasted in an experiment. And in contrast to novices, experts rate the similarity between wines along established wine categories (Chardonnays and Muscadets). The experts' explicit knowledge allows them to more easily attend to, categorize, and then remember *meaningful* constellations of flavor: flavors that go together because of grape or production processes. Without this explicit knowledge and taste training, even the frequent wine drinker is likely to miss flavor subtleties in a vintage.

And this is an explanation with which Master Sommelier Steven Poe fully agrees. "There's no question that the more knowledge you have, the more your tasting is informed. If you know, for example, that they use tufa subsoil in the Anjou-Saumur [France], then when you're trying a cabernet and getting that chalky, stony minerality on the wine, you become confident that you're tasting a Cabernet Franc from the Loire Valley, probably Chinon. And knowing this can draw your attention to other tastes associated with that region."

Thus, having explicit knowledge about wine production allows Poe to more effectively attend to meaningful configurations of sensory dimensions. In fact, this conceptual advantage is shared with many other types of experts. It's long been known that chess experts can be given a quick look at a chessboard from an in-progress game, close their eyes, and then recite the positions of all the pieces. It turns out, however, that this skill is dependent on the pieces being in a configuration derived from an actual game and the rules of chess. If the pieces are randomly placed on the board, chess experts are no better than novices at remembering positions. The same types of effects have been shown for master bridge players, volleyball and basketball coaches, as well as transcribers of musical notation and circuit diagrams. And the same is true of wine experts. In memory experiments, wine experts are better able to remember palate-related word pairings when both words pertain to a single wine type ("full-bodied," "prune"; both descriptions for syrah). However, when the words are *not* paired in a way that refers to a single type of wine ("full-bodied, "lime"), then experts are no better than novices in remembering the pairs. Wine expertise, like the other types, is dependent on remembering component clusters that are *meaningfully* arranged.

And the expert's conceptual and language skills with wine seem to off-set a failing of language and memory that plagues most of us. Typically, we think of human language as providing an advantage in remembering things. And there is no doubt that naming things helps us remember a list of items (shopping lists). But there are situations in which using language not only fails to help memory but also makes things worse. Consider this scenario: You've just witnessed a petty crime and have gotten a good look at the perpetrator. You feel confident that you would be able to identify the perpetrator from a photograph and you tell a police officer this. Before seeing a lineup, however, the officer asks you to provide a detailed description of the perpetrator's face. You do your best with this: shifty eyes, pug nose, crooked mouth. After completing this description you see a series of photographs, one of which, you are told, is of the likely perpetrator. But despite your earlier confidence, you find it surprisingly difficult to identify the perpetrator from the photos. The police officer tells you, yeah, it happens to everyone.

It turns out you'd be substantially better at identifying the perpetrator if you had *not* first provided a verbal description of him. Your verbal description interferes with your simple ability to recognize the face. This phenomenon, known as *verbal overshadowing,* is well known in the perceptual literature and occurs for many types of stimuli including visual images, colors, and music. It's thought to occur whenever your perceptual skills substantially exceed your verbal skills in a domain. You are typically superb at recognizing faces, as you'll learn in Chapter 9. You are much, much worse at describing faces. When forced to describe a face verbally, it distracts you from the important visual information you use to recognize faces—information that is much harder to verbalize.

Verbal overshadowing occurs for smells and tastes as well. As you've learned, smells and tastes are notoriously difficult to describe in words, so that when you are asked to do so, it can distract from whatever non-verbal sensory information you would normally use for recognition. And this seems especially true for tasting wine. Regular wine drinkers (two glasses a week) are fairly good at recognizing a wine they had tasted five minutes earlier (when given a choice of four). However, if they are asked to provide a detailed verbal description of the wine before attempting to recognize it, they become *completely unable* to recognize the target wine from its taste. Their descriptions verbally overshadow whatever information they would normally use to recognize the wine's flavor.

But verbal overshadowing doesn't affect sommeliers. Their training and explicit knowledge allows them to avoid it. When experts are asked to recognize a wine they just tasted (when given a choice of four), they are, unsurprisingly, quite accurate. But even if they are asked to provide a verbal description of the wine after first tasting it, their recognition skills are not at all impaired. An expert's verbal descriptions likely capture just the type of conceptual information they use in attending to the meaningful configurations of a wine's component flavors.

Wine expertise has become one of the most studied forms of perceptual expertise. This is likely based on a number of factors including the extensive formal training involved in attaining expertise, as well as the impressive subtleties the expert is able to discern from a sense modality most of us feel provides little insight. But all evidence suggests that wine experts are not sensorially different: they show no greater sensitivity in their tongues and noses, as such. Instead, the wine expert's skills are based on learned conceptual knowledge, attention to flavor configurations, and language use.

These traits are fully consistent with the aforementioned neuro-physiological differences revealed by brain imaging research. Sommelier brains likely go through changes induced by their extensive training and analytical approach to tasting. And it is no doubt the brain's inherent neuroplasticity that allows these changes to occur. In this sense, wine expertise is attainable for all of us. With serious studying of wine theory and fundamentals, along with practice in tasting, your brain could start lighting up like the brain of a master sommelier.

IT TURNS OUT THAT what happens on your tongue is only a small part of how you taste food. The way something tastes is a result of myriad influences from what you smell, touch, see, and even hear. What you taste also depends on what you expect and what you know. Your tasting brain is like a kid with a new chemistry set, trying all combinations of information associated with food to determine the most exciting and appetizing taste experiences.

Part IV

Touching

Blind painter John Bramblitt first draws the outline of his subjects with a raised-line paint that he later touches to determine where to fill in colors.

CHAPTER 6

Rubber Hands and Rubber Brains

JOHN BRAMBLITT DIDN'T START PAINTING UNTIL HE LOST HIS SIGHT. It was a difficult time. Bramblitt was in his late twenties and was unaware that his sight was seriously degrading until he was sideswiped by an unseen car. He was also worried about having the severe epileptic seizures that had already taken their toll on the visual regions of his brain. And he was angry. In fact, he believes that taking up painting after losing his sight was mostly an act of defiance.

Bramblitt had sketched casually in his youth, but he had no experience painting on canvas. Learning to do so as a newly blind individual was painstaking. While his 25 years of visual experience provided him with mental images of what he wanted to paint, he was uncertain how to render these images on a canvas he couldn't see. He knew that he would need to use his sense of touch to create the images, but was unsure how paint could allow this. Then he discovered "puffy paint." Puffy paint is typically used for decorating fabric and leaves a thin, raised line—a line Bramblitt can touch. Puffy paint dries almost instantaneously so that it can be touched just after it is applied. This allowed Bramblitt to develop a technique in which he first uses the puffy paint to produce a detailed drawing of his subject on the canvas. He draws with the tip of a small paint bottle held in his right hand and carefully uses the fingertips of his left hand to determine where he has placed each line. This allows him to guide the position of the next line he'll lay down. Initially, Bramblitt found this to be a very slow process. But he has now developed his sense of touch and interhand coordination to the degree that he can draw at a speed not much slower than that of a sighted person.

Once the puffy line drawing is complete, he fills in the areas between the lines with different colors of oil paint. This method involves feeling his way across the canvas with his left hand, as he paints with a brush held in his right. His choice to work with oils has been critical to the process. While oil paint is messier, more pungent, and dries much slower than acrylics, it offers something that no other paint can: idiosyncratic viscosity. Bramblitt has discovered that different colors of oil paint feel different based on the ingredients used for their pigments. "Depending on the manufacturer, white feels thicker on my fingers, almost like toothpaste, and black feels slicker and thinner. To mix a gray, I'll try to get the paint to have a feel of medium viscosity." And by sticking with a single manufacturer for each of his paints, he is able to recognize and mix all the other colors he uses in this way.

Bramblitt typically paints at least eight hours a day and has done so for nearly 10 years. This intensive experience has allowed him to fine-tune his touch skills so that he can work more efficiently. These days, he doesn't need to touch the paint with his fingers so much. Instead, he is able to get a good sense of what paints and mixtures he's using by how the brush feels in his hand as he pulls its bristles through the paint. The viscosity of the paint creates a resistance in the bristles that gets transmitted to the brush's handle. He can feel the paint's viscosity, by proxy, through the brush. Similarly, as he paints on the canvas, he is now able to feel the raised lines of the puffy paint with his brush. His awareness of the resistance to his brush strokes allows him to keep his colors within the lines' borders. This frees up his left hand to take in more of the raised-line drawing so he can anticipate what parts he'll paint next.

The subjects of Bramblitt's paintings have broadened since he began. While he still paints subjects from memory—familiar landscapes, his old church, famous musicians—he also paints still lifes and portraits of his friends. For these latter subjects, he first spends a good amount of time feeling what he'll paint: a chair, wine bottles, a friend's face. And here he believes that he may have an advantage over sighted painters. "When you touch something to get its detail, it's a much more active experience than looking at something. When I touch a chair in order to paint it, I become aware of all these details a sighted person could miss: how the wood grain on its seat differs from that on its legs; the type of screws that hold it together. And touching a face gives me even more. When you touch a face, you get an immediate connection with that person. That connection helps me choose the colors I'll use in the painting—that and how similar their skin texture feels to the different oil paints I use." And the colors are the first thing one notices about Bramblitt's work. While the subjects of his paintings are immediately recognizable, proportioned, and smartly stylized, the colors are supremely vibrant and nearly psychedelic in their rendering.

Bramblitt makes a good argument for considering painting as more than a visual medium. He has taught his style of painting to both blind and sighted students. For sighted art students, he often asks them to first draw an object from sight—an apple perhaps. The outcome is what one would expect, drawings of apples that are immediately recognizable, but quite literal and homogeneous. He then asks students to close their eyes and touch the apple for a minute. He tells the students to then open their eyes and paint the apple they felt. Reportedly, the resultant drawings are very different from one another and more aesthetically interesting.

And Bramblitt has developed an appreciation for the work of other artists. Through the invitation of a private collector, he has had the opportunity to touch original paintings by Cézanne, Rembrandt, and Van Gogh. While he says that most oil paint dries to an indistinguishable consistency, he is able to feel the brush strokes in the paint. "When you touch a painting, you can almost get in the head of the painter. When I feel long languid brush strokes, I know that the painter must have been in a very calm state. Shorter, more abrupt, inconsistent strokes probably means that the artist was in a more excited state when working on that part of the painting." This experience, along with having a sighted person describe the image, may give him a more intimate appreciation for a painting than most sighted people have.

These days Bramblitt is a much happier person. He has married a fellow artist and has a young child. His art is selling well, both from his Web site and gallery exhibitions. He gets great pleasure from sharing his techniques and experience with others, and plans to get a master's degree to teach art at the college level. And he continues to get tremendous satisfaction from the eight hours a day he devotes to his unique style of painting.

■ ■ ■ ■ ■

JOHN BRAMBLITT HAS DEVELOPED his touch skills in particularly impressive ways. But the enhancement of the touch sense is known to generally occur for blind individuals. This may not seem surprising for blind individuals who use Braille, the raised dot patterns allowing the blind to read with their fingers. But in fact, only about 10 percent of the nation's blind currently use Braille—a consequence of mainstreaming blind children into public schools as well as the ubiquity of audio translators for any text material accessible by computer. But even blind individuals who have no Braille experience, and have been blind only two years show greater touch skills than the rest of us. In fact, a blind individual's touch sensitivity seems to counteract the natural decline of aging and is typically as great as that of a sighted person 25 years younger.

Why do blind individuals have enhanced touch sensitivity? The answer lies with one of the most important perceptual brain discoveries made over the last 20 years: *cross-modal plasticity*. As intimated in earlier chapters, when faced with sensory loss, we often show compensatory enhancement of our other senses. This likely helps Dan Kish mountain bike, Houston Heat players stop a beeping baseball, and John Bramblitt feel his way through a painting. But not until recently have scientists discovered just how sensory compensation occurs in the brain. And for blind individuals, brain areas typically dedicated to visual perception (occipital lobe) often become co-opted by the auditory and, particularly, the touch sense.

When you touch something, you activate a strip of your brain that runs from a spot just above your ears all the way to the top of your head (somatosensory cortex). Touching also activates higher-level brain areas needed for recognition and reaction. However, with blindness, touch also strongly activates *visual* brain areas. And it seems that activation of visual regions might be integral to the enhanced touch skills of the blind.

There are numerous reports of experienced Braille readers who lose the ability to read Braille as a result of stroke-induced damage in their visual brain areas. And a recently developed technique to induce "transient lesions" shows similar effects. This technique (transcranial magnetic stimulation) uses an external device that is held to the head and sends brief but potent magnetic pulses to specific brain regions. These pulses render the target brain area temporarily unusable. The technique is considered safe and only affects the brain while the pulses are administered. When using this device on the touch strip area of the brain (somatosensory cortex), both blind and sighted individuals find it difficult to recognize even simple shapes from touch. When applied to the "visual" brain regions, however, sighted individuals show no change in their touch abilities. In contrast, blind individuals find it nearly impossible to recognize complex patterns from touch and many also experience distorted felt patterns. These findings show that activation of visual brain areas is an important part of the blind's touch perception skills.

This type of cross-modal plasticity occurs in individuals who have been sightless for many or just a few years, as well as for those who do or do not use Braille. It seems that an unused visual brain is ready to begin reassigning itself to the service of touch as soon as it's needed. And as you'll soon learn, hardly any visual deprivation is required for the visual brain to help touch.

YOUR VISUAL BRAIN IS READY TO HELP YOU TOUCH

IMAGINE YOURSELF IN THIS EXPERIMENT. You don a specialized blindfold that prevents your eyes from receiving any light. You then check into a hospital room where you'll live blindfolded for the next five days. You are supervised by nurses and researchers, but being in a new environment without the benefit of sight does take some getting used to—as do the multiple tests you will be subjected to during the five days. Early on the first day, you are placed into a large brain scanner and asked to touch a series of raised-dot arrays with your fingers. Your task is to determine whether these dot patterns, presented sequentially in pairs, are the same or different. After this initial test, you begin your first six-hour Braille lesson, a lesson that will be repeated over the next four days.

Incidentally, by the second day you start experiencing a number of side-effects from the continuous blindfolding. You have visual hallucinations of both amorphous and recognizable images, as well as an initial dullness of flavor (see Chapter 5) and oversensitivity to temperature and sound.

Then on day five, before removal of your blindfold, you are subjected to another brain scan as you again attempt to discriminate dot patterns. You no doubt notice that the dot pattern task now seems easier than it did on day one. But before you get cocky about your improvement, you are subjected to one last test that immediately renders you incompetent with the task. For this last test, a device is held over the back of your head that makes periodic clicks as you try to match the dot patterns. Strangely, you find the task nearly impossible and have trouble determining the patterns altogether.

THESE WOULD NOT BE the five most enjoyable days of your life, but they might be the most scientifically productive. Five days of blindfolding would change you and your brain in fascinating ways. Most obvious would be your substantial improvement in discriminating the dot characters: the intensive Braille training would seem to have helped. But you may be disappointed to learn that you would have improved these touch skills even without the Braille training. Five days of *blindfolding alone* can enhance basic touch skills. Also, regardless of whether you had the Braille training, your brain would have changed over the five days of blindfolding. The last brain scan would reveal that when you now touch complex patterns, your visual brain is activated in a way similar to that of an individual who is truly blind. But for participants who were not blindfolded, these brain changes would not occur, even if they did have the intensive Braille training. Five days of visual deprivation are enough to establish visual brain involvement in touch tasks, as well as the task performance advantages that involvement provides.

As extra support for this conclusion, remember the last (imaginary) test you took when you were blindfolded? The one where a clicking device held to the back of your head completely disrupted your touch skills? That device induced the infamous transient lesion in your visual brain area. So while the classic touch areas of your brain were unaffected, disrupting your visual brain disrupted your new touch skills after the blindfolding period. This confirms that your touch skills were enhanced by involvement of your visual brain—an involvement enabled by the blindfolding procedure. The visual deprivation allowed your plastic brain to become better at touch.

■   ■   ■   ■   ■

IF FIVE DAYS OF BLINDFOLDING seem an uncomfortable way to induce cross-modal plasticity, perhaps 90 minutes are more appealing. That's all it takes to enlist visual brain involvement in a pattern touching task, even if the blindfolded period doesn't involve touch training of any kind. Ninety minutes in the dark are also enough to increase your touch sensitivity to the orientation of a shallow plastic grid, as well as the presence of a small gap in a plastic surface, and the identity of barely raised plastic letters. Again, no touch practice is necessary for this enhancement: the short-term deprivation of vision is enough to enlist your visual brain for the touch cause.

These findings have led many neuroscientists to believe that cross-modal plasticity is a general strategy of all brains, regardless of the sensory ability of the owner. Many researchers now think that if given the chance, the visual areas of our brains are willing and able to help with much more than just vision. Cross-modal plasticity is not something exhibited only in sensory-deprived individuals. It is a part of your brain's strategies, ready to help out in many everyday contexts.

But before running out to buy a blindfold and checking into the hospital, you should know some additional facts about the cross-modal plasticity induced by short-term visual deprivation. First, the touch enhancement induced by temporary blindfolding is usually short-lived. Even after the five days of darkness, the touch system usually goes back to its original proficiency within 24 hours, and the visual brain loses much of its involvement with these touch tasks. (The issue of short-term versus long-term plasticity will be discussed later.) Second, there are other ways to get your visual brain involved with touching that don't involve blindfolds.

It's likely that your visual brain already helps your touch system every now and then. Research suggests that this help is restricted to specific touch tasks and is not nearly as great as what the visual brain offers the blind, or even the blindfolded. But recent experiments using brain scanning and virtual brain lesion techniques show that your visual brain can help out when you normally touch something to determine its *spatial* detail. This may help you use some form of visual imagery when recognizing spatially complex objects through touch. You may use visual imagery when feeling for a particular pen in your purse or fumbling for your house key in the dark.

Regardless of why your visual brain is recruited for complex touch tasks, and regardless of the fact that this recruitment is less than that for the blind or even blindfolded, your visual brain is ready for touch when it's needed.

You Can Taste Scenes

You've been blindfolded and asked to stick out your tongue. You oblige, and the researcher gently lays a postage stamp–sized circuit board on your tongue. There is a thin flat cable leaving the board, coursing over your bottom lip and connecting to a box hanging from a string around your neck. Also connected to this box is a small camera that has been fastened to your forehead with a headband. The researcher then switches on the box and you feel a tickling sensation on your tongue, not unlike the fizz of champagne bubbles. Then, despite the blindfold, the researcher asks what you see in front of you.

This device is the Tongue Display Unit, or TDU, and has been designed to allow visually impaired individuals to, essentially, see with their tongue. It is the most recent version of a long line of tactile aid devices that have been designed since the advent of the television camera. Because the touch sense can be developed to recover such spatial detail, especially in blind individuals, it makes a good channel around which to design visual substitution devices. One of the first systems developed in the 1960s was composed of a video camera that translated its images to an array of 400 small vibrating posts placed against the back of a user. The brighter parts of the images were translated into faster vibrations in corresponding posts in the array. The user would feel the pattern of vibrating posts on his or her back and attempt to identify what the camera saw.

Without any training, subjects would be able to use this system to discriminate horizontal, vertical, and curved lines, as well as to detect simple targets so as to both point and turn toward them. With a moderate degree of training, users would be able to recognize geometric and face shapes, and simple scenes. They'd also be able to follow the movements of a rolling ball well enough to intercept it as it falls off a table. And a very recent demonstration suggests that users can learn to remotely navigate a robot through a real 3-D maze based on this touch stimulation array.

One of the most interesting aspects of touch-based vision substitution systems is how a user's experience of the stimulation changes over time. Users initially experience the vibration patterns on their backs as just that, and they report consciously analyzing these patterns to create a mental image of the target objects. But after many hours of training, users report experiencing the objects as existing *outside* their bodies, in specific locations in front of them. Thus, with experience, the skin ultimately becomes like the eyes in experiencing stimulation as out in the world, rather than on the body. Also with practice, it doesn't matter where the vibrating post array is applied: it can be moved from a user's back to the legs or arms without any decline in the user's ability to recognize objects. It's as if the brain learns to interpret the skin as a unified sense organ that detects patterns of touch. These touch patterns allow us to functionally see with our skin.

Most recently, touch-based vision substitution systems have moved to the tongue, of all places (the TDU). The tongue provides a good location for a touch array because its receptors are especially close to the skin surface, and its constant moisture provides excellent electrical conduction. This allows for the array to be composed of small, low-voltage electrical transducers instead of the energy-intensive vibrating posts. A lower-energy requirement also allows the system to be much more portable. In addition, the touch brain devotes an especially large amount of its space to tongue stimulation. This means that touch resolution at the tongue is greater than for most of the external skin.

To use the TDU, the array is placed on your tongue, and the transducers transmit a very low electrical current to your tongue's surface in the pattern captured by the video camera on your head. After a few minutes, you can recognize shapes, numbers, and the orientation of eye-test *E*'s. With practice, you can use the TDU to navigate cluttered environments: you essentially taste scenes. As with the earlier vibrating-post systems, practice with the TDU allows you to experience the perceived images as external to your body. And very recent research shows that when blind users have some practice with the system, their visual brain regions are activated by the TDU.

As of this writing, the "image resolution" on the tongue or skin for even the best touch-based systems is equivalent to that of a very unfocused camera lens. It's unclear whether these systems will ever posses the image quality and convenience to take the place of canes and guide dogs. Still, the fact that even systems with poor resolution can allow you to recognize some objects and traverse some environments evidences the spatial sophistication of your touch system and its underlying neuroplasticity.

THE MORE YOU TOUCH, THE MORE YOUR BRAIN CHANGES

PLASTICITY DOESN'T JUST OCCUR between your perceptual systems but also *within* a single perceptual system. Short-term and long-term experience can influence changes in your perceptual nervous system, enabling improvements in a perceptual skill. And for no other sense do we understand more about perceptual plasticity than for your touch system.

Intensive practice with touch can change the organization of your brain's touch areas. For example, not only do the visual brain areas change in the blind, but the brain's *touch areas* (somatosensory cortex) also change, especially with Braille experience. Most Braille users are taught to touch the dot characters using the three middle fingers of each hand. This allows for touch "glances" similar to those you use to visually read words and their parts as you scan lines of text. It turns out that the amount of touch strip area dedicated to the Braille reader's middle three fingers is disproportionately larger than it is for sighted, non-Braille readers. For Braille readers, these touch brain areas also respond more quickly to touch stimulation.

A similar reorganization occurs for experienced musicians. For violin players, the touch brain's mapped space for their left-hand fingers—the ones used to dexterously press the strings down on the violin's neck—is substantially larger than the mapped space for the right-hand fingers (used to simply grip the bow). Further, the magnitude of this left-hand/right-hand difference in mapped space is greater for violinists who've played for longer periods of time.

And a musician's touch experience enhances his or her touch sensitivity. Professional pianists have much better "two-point" touch sensitivity than nonmusician controls. Two-point sensitivity is a standard measure of touch used in both research and for clinical diagnosis. The method involves placing two blunt needles together onto a fingertip, and then slowly moving them apart (each time removing and reapplying the needles) until the subject is aware of feeling two separate needles. It turns out that professional pianists need less distance between the needles before they can feel both. Further, this two-point sensitivity increases with the number of hours per day the pianist practices. So being a musician not only tunes your ear, it also tunes your skin and its associated brain regions.

But you don't need to be a musician to induce touch-based neuroplasticity. Neuroplasticity likely occurs whenever you use a tool for as little as 10 minutes. Imagine being given a set of tongs (long chopsticks) to manipulate the position of a rubber cork between three different targets marked on a wooden board. Ten minutes of this task would be enough to change the organization of your touch brain (somatosensory cortex). From using the tongs, the representations of your thumb and pinky fingers would diverge on your brain map as if your brain needed to accommodate the prongs into its mapped space. As far as your touch brain is concerned, the tongs have temporarily become an extension of your hand. Something similar could very well happen whenever you intensively use pliers, a screwdriver, hammer, or even a pencil for a short period of time.

But if you don't feel like picking up a tool, there are even easier ways to induce neuroplasticity and enhance your touch sensitivity. In fact with the right equipment, you don't need to do anything at all.

In the touch stimulation method known as *coactivation*, an untrained subject is lightly tapped on the fingertip by a small (one-third-inch) circular probe tip. The tapping occurs with a random rhythm averaging about one tap per second, and continues over the course of hours or even days (the tapping device is portable and unobtrusive). While small, the one-third-inch size of the tap is large enough to *coactivate* multiple touch nerves on the fingertip, which turns out to be critical for the method to work.

This simple device is enough to change your brain and enhance your sensitivity to touch. With three hours of coactivation stimulation, the area of your touch brain strip dedicated to that finger expands by measurable amounts. In turn, your two-point sensitivity improves, and you would be better able to determine the orientation of small plastic gratings with that fingertip. In fact, your improvement in touch sensitivity would be directly proportional to the degree of reorganization in your brain.

This coactivation method has been used with the elderly. One study showed that after three hours of stimulation, the two-point sensitivity of 80-year-olds improved to levels comparable to that of fifty-year-olds. Brain scanning of these subjects showed the predicted reorganization in the touch strip, with more area dedicated to the stimulated finger. It seems that the well-known decline of the elderly's touch sensitivity is, to some degree, reversible and that neuroplasticity continues to be active in the aging brain.

But again, before you go out and buy a finger coactivation device, you should know how long its effects last. With three hours of coactivation stimulation, your enhanced sensitivity, and associated brain reorganization, last about six hours. With three days, the changes last an additional 48 hours. There are some ways to modify coactivation to lengthen the effects. Applying coactivation for three hours to *all* the fingers of one hand lengthens the effects to beyond 24 hours. And if the coactivation tapping is much more rapid—20 taps per second— then just 20 minutes of the procedure can induce sensitivity enhancements lasting up to two days.

Clearly though, coactivation-induced changes do not have nearly the longevity of the enhanced touch sensitivity shown by musicians and Braille users. And it is here where the concepts of short-term and long-term plasticity become relevant.

Your Brain Uses Short-Term Plasticity to Establish Long-Term Plasticity

With short periods of coactivation touch, your touch brain goes through a temporary period of reorganization. This *short-term plasticity* is a result of nerve cells using connections they haven't used before, or at least not very often. With coactivation or other types of short-term touch training, touch brain cells start communicating with neighboring cells with which they've had dormant connections (this process is called *unmasking*). It's as if touch experience opens the doors between neighboring cells so that their influences can be more widely shared. This extended sharing of cell activity causes the reorganization of the touch brain's map to emphasize a stimulated fingertip. It also underlies the visual brain's involvement in touch under conditions of blindfolding. But once these short-term manipulations end—the coactivation ceases, the blindfold is removed—the doors between the cells begin to close, and the extended activation of neighboring cells wanes.

With enough repetitive touch stimulation (or sustained visual deprivation), however, the doors between touch brain cells begin to open more readily and new doors are actually constructed. This long-term plasticity is thought to arise from the sustained effects of short-term plasticity. When the extended sharing of cell activity is sustained and repeated over long periods of time, the cells start finding new ways to maintain this sharing. This can involve structural changes in the cells themselves, so that they sprout new limbs that reach out to neighboring cells. These limbs then provide an additional doorway to the neighboring cell, making extended activation a near certainty. The new growth establishes more permanent interactions between cells, allowing the once-dormant connections to be active full time. And it is this process that establishes the type of long-term changes observed in musicians and Braille users. Put simply, if the effects of short-term plasticity are sufficiently sustained, long-term plasticity is established.

To get a sense of how short-term plasticity changes to longer-term plasticity, consider this experiment. The experiment examines changes in the brain's map for touch *motor control*, rather than touch perception, but the same brain strategies are at work. A group of nonmusicians were taught to use their right hand to play a complex pattern on a digital piano keyboard. The keyboard was connected to a computer so that the subjects' fingering errors, fluidity, and rhythmic consistency could be monitored. Subjects practiced this pattern for two hours on each of five consecutive days. Unsurprisingly, the precision and consistency of the subjects' playing increased significantly over this period.

Brain scans (mapping, actually) were conducted on each day before and after the practice sessions. These scans revealed that across the five days, the brain areas dedicated to controlling the muscles of the right-hand fingers increased. However, this increase only appeared *after* the two-hour practice sessions on each day. Scans conducted before the subjects practiced failed to reveal any change in the size of the relevant brain areas. The practice-induced plasticity was short-lived, reverting to its original state by the next day. At the same time, however, once a practice session reestablished the plasticity changes, those changes did benefit from the accumulated experience. This fact, no doubt, was related to the improved performance subjects displayed over the five days.

Now, at this point in the experiment, half the subjects were let off the hook: they stopped practicing the keyboard pattern altogether. The other group, however, continued practicing the pattern for two hours, Monday through Friday, for the next *four weeks*. On each Monday morning and Friday afternoon, all subjects' brains were scanned. The results were stunning. While the group of subjects who stopped practicing showed brain area maps just like those they exhibited before the experiment began, the practicing group showed continued reorganization. But most interesting was the fact that for the group that had practiced for four weeks, their Monday morning scans—scans taken after a weekend *without* practice—*also showed this reorganization*. This means that the reorganization induced by four weeks of practice was more permanent, evident even after two days of rest. It's likely that the four weeks of practice created the longer-term neuroplasticity indicative of actual nerve cell growth.

The notion that short-term plasticity is a precursor to long-term brain changes likely applies to all of your sensory systems. Sound engineer Jay Patterson, coffee smell expert JC Ho, and flavor tester Linda Blade likely all went through short- and then long-term plasticity as they established their skills. Practice doesn't just make perfect: it makes plasticity permanent.

But not all neuroplasticity has positive effects.

YOUR BRAIN REMEMBERS THE TOUCH YOU'VE LOST

TALKING WITH BEN CURRY, you quickly realize you're with a man who's been mourning. He speaks quietly and slowly, and he looks down often. And Curry, who is a slight man of 55, talks openly about his sadness. "I've been grieving more than when I lost my mother, father, and brother. I've never felt as much grief as I have for my leg. It wasn't until six weeks after my amputation that I was even able to look at my naked stump and clean it myself." Even without looking at his stump, he's constantly reminded of his loss through a phenomenon that afflicts 90 percent of amputees: *phantom pain*. Phantom pain is the sometimes intense discomfort reported by amputees in what they experience to be their missing limb. Curry's pain is moderate, and is often experienced as an itching and burning in his missing big toe, the toe that originally led to his amputation.

Curry lost his left leg after accidentally cutting his toe. Because of his diabetes-related neuropathy, he was unaware of the extent of his injury until gangrene had set in. The gangrene spread to his lower leg, requiring amputation from about two inches below his knee. Two weeks after his amputation, he started to feel tingling in the missing toe. (Phantom pain is often experienced at the injured site.) "At first it felt as if my missing toe was asleep. It would come and go throughout the day, and it would last for about an hour each time. But after a week or so, the tingling would come with a mild burning pain, and itching. And the itching can be awful: it's like a bad case of athlete's foot. In fact, I'm feeling it now." Curry reaches down to the empty space where his foot and toe would normally be and starts rubbing and scratching the air. He lifts his head and smiles slightly, "In a strange way, itching my phantom toe feels good."

For decades phantom limb pain was thought to be caused by nerve damage in the remaining stump, or psychosomatic symptoms induced by the emotional trauma of losing a limb and dealing with its consequences. But for about 10 years now, scientists have largely believed that phantom pain is a result of overexuberant neuroplasticity. Losing an appendage induces a reorganization of the brain's touch centers. Typically, a remaining body part whose touch brain representation is adjacent to that of the missing limb takes over the vacated brain area. But because of the massive change induced by amputation, this plasticity rarely goes smoothly.

In fact, individuals with constant and intense phantom pain show a more extreme reassignment of their brain's touch regions. And when this reassignment involves more distant parts of the body, phantom pain can be especially bad. For example, intense phantom pain seems to be particularly common in individuals whose hand amputations have allowed vacated brain regions to be reassigned to areas of the *face*.

While the face and hand are separated in the body's space, they are actually represented in adjacent areas on the touch brain strip. If, after amputation, the face map merges into the regions that once mapped the hand, it is likely that the amputee will report intense phantom pain in the missing hand.

Just before his amputation, Ben Curry was told of the likelihood of phantom sensation and pain. He was also told that it would be a consequence of his brain trying to realign itself. And this is actually one of the reasons he scratches and rubs his missing foot. "I'll let my brain think that I'm going along with it. It's like my brain is looking for my missing foot. It gives me a phantom sensation that is so vivid, I can go down to where my toe used to be and scratch exactly that spot. Of course, I can't feel my hand touching my phantom foot. Really, I just scratch to patronize my brain."

Phantom pain and sensation can last anywhere from a few months to a lifetime, depending on its severity and location. But the recent insights about the neuroplasticity basis of phantom pain have motivated new treatment options. Phantom pain has historically been treated with pain medication or additional surgery to remove the nerve endings at the stump, an extreme and not particularly effective option. Newer options making use of neuroplasticity have involved remapping the touch brain back to the remaining stump by direct stimulation of its nerve cells. Other treatments take advantage of the visual effects on the touch brain's neuroplasticity—effects you will learn about shortly. In one treatment, mirrors are positioned to reflect a view of the remaining limb to the patient, so that the reflection looks like the limb that has been amputated. The remaining limb is then visibly touched so that its reflection can fool the brain into interpreting the touch stimulation as *at* the missing limb. With repeated sessions, mirror treatment has shown some success at treating phantom pain.

Curry is slowly emerging from his grief. He's more optimistic about his future: "You may laugh, but I have plans to run a marathon after I get my prosthetic. I've always been a runner, and I know that I will be again, even if it takes a while." And he seems to have a healthy perspective on his phantom pain. "You just have to know that it's your brain's way of dealing with the loss. You need to help your brain go through this." Again,

Curry bends down to scratch his phantom foot and I get the sense that he's taking care of more than just his touch brain.

NEUROPLASTICITY DEPENDS ON WHAT YOU *THINK*

VERY RECENT RESEARCH IS showing that neuroplasticity in the touch areas of your brain can be brought about in more subtle ways than being touched repeatedly on the finger or through practice with Braille or playing a keyboard. In fact, neuroplasticity can be induced with no touch at all.

Recall the experiment in which subjects were taught to play a repetitive pattern on a keyboard. There was actually one more group of subjects tested in that experiment. This group learned the keyboard pattern on the first day, but then didn't play again during the four following days. Instead, these subjects *mentally rehearsed* the pattern on each of the remaining days. They were asked to imagine their fingers playing the pattern on the keyboard repetitively for the two-hour sessions. After four days of mental practice, this group was able to play with more accuracy and fluidity than a control group that learned the pattern but didn't practice (physically or mentally). (The mentally practicing group's performance was not quite as good as the physically practicing group, however.) These performance results are similar to other findings on mental practice. Intensive mental practice has been shown to enhance performance on a wide range of skills, from the musical to the sporting.

Most astonishing, however, were the results of the brain scans. It turns out that the mentally practicing subjects showed the *same* neuroplasticity as the subjects who physically practiced the pattern: their brains shifted to dedicate more area to the right-hand fingers. Imaginary practice, like real practice, can change your brain.

In fact, if you happen to be a practitioner of Tai Chi, chances are you've already imagined yourself to better touch sensitivity. Tai Chi is a martial art typically used for meditative exercise. It involves slow, precisely controlled movements of the arms, hands, legs, and torso. These movements do not involve contact with another person or (usually) one's own body. However, serious Tai Chi practitioners devote substantial mental energy focusing on the hands and, particularly, the fingertips. Focusing on the fingertips is thought to help the practitioner concentrate on the entire body, from its core to its extremities.

This unique aspect of Tai Chi provides a useful way to evaluate the effects of mental practice on touch sensitivity. It turns out that experienced Tai Chi practitioners show enhanced fingertip sensitivity when asked to judge, from touch, the orientation of shallow plastic grids. Further, Tai Chi–related enhancement is more pronounced for older practitioners, showing again that plasticity is active in the aged. So, just *thinking* about your fingers in a repetitive, disciplined manner can improve their sensitivity.

And how you *attend* to touching can influence your touch sensitivity and its underlying plasticity. Research shows that concentrating on different aspects of the exact same touch stimulation can determine how your brain reorganizes.

NEUROPLASTICITY CAN BE INDUCED BY A PRETEND LIMB

PERHAPS THE MOST COLORFUL examples of how what you think can influence your touch plasticity are provided by *rubber hand illusions*. To truly appreciate the oddity of these effects, imagine yourself in this experiment. You are asked to sit at a table with your forearms facing down against the tabletop and at about shoulder-width distance from each other. Your hands are in a relaxed position, with your fingertips touching the table. A large wooden divider is then placed just to the right of your left arm so that you can no longer see that arm. In its place, a rubber left arm with hand is positioned in front of you so that it looks like it could be your own. To enhance the effect, a large cloth is spread over your right arm and the rubber arm, so you see only your own right hand and the rubber left hand in front of you.

Now comes the really bizarre part. The experimenter takes two small paintbrushes and starts simultaneously stroking the index finger of the rubber hand along with the same digit on your unseen left hand. After about two and a half minutes of this, odd things happen. You start experiencing the rubber hand as your own. You may comment that the rubber hand feels like it has become part of your body and describe the experience as "bizarre," "scary," or "trippy," as subjects often do. If the lights were then turned off and you were asked to indicate the position of your *real* (hidden) left hand relative to an illuminated ruler, you would choose a position somewhere between your real and rubber hands.

Not only would you have the perceptual experience that the rubber hand has become your own, but your body would also respond as if it were true. This can be demonstrated in a number of ways, many of which are, shall we say, sadistic. Imagine this. After you tell the experimenter you experience the rubber hand as your own, he quickly takes the index finger of your rubber hand and violently bends it backward. Or just when you least expect it, the experimenter abruptly plunges a syringe deep into the top of your rubber hand. Both evil acts would have the same effect. You might jump, grimace, or shriek, but more importantly, your autonomic nervous system would react as if you were truly threatened. Your heart rate would speed and your skin would start sweating, creating a measurable electrodermal response. And your brain would react as if it were in danger. Areas associated with anxiety, pain anticipation, and motor reaction (insula, anterior cingulate cortex, premotor areas) would all respond.

But such sadistic methods are really unnecessary. More peaceful measures can be used to show how your body adopts your rubber hand. After 45 minutes of watching your rubber hand being stroked (in synchrony with the felt stroking of your unseen left hand), brain scans would show that your touch brain reorganized in an idiosyncratic way. Critically, this reorganization would not occur if the seen stroking were performed *asynchronously* with the felt stroking: a manipulation that would also eliminate the feeling that the rubber hand is your own.

And with minor modifications of the rubber hand method, your touch brain could be made to reorganize in fascinating ways. Imagine looking down at your rubber hand and realizing that it is now connected to a rubber arm that is twice its normal length. Essentially you've grown a rubber baboon arm. If you now saw your rubber hand stroked as you felt a synchronous stroking on your real, unseen left hand, your touch brain would reorganize to incorporate your long baboon arm. Specifically, your brain would now represent your fingers in a smaller area than it would if you were watching stroking of a rubber hand connected to a normal length arm. It's speculated that the touch brain space for your hand shrinks to accommodate the increased touch space now needed for your long baboon arm. In contrast, the opposite brain reorganization occurs if your own stroked hand appears (through a video monitor) to be twice its normal size (your touch brain would represent your fingers as farther apart than normal). So, you adopt your new rubber hand to the extent that your touch brain reorganizes to accommodate its specific characteristics.

But it seems you only can deal with two hands at a time. As your body adopts your new rubber hand, it shows signs of disowning your real, left hand. Your unseen left hand's reactions to touch start to slow. For you to feel that your two real hands are being touched simultaneously, your disowned left hand would need to be touched sooner than your right. To make matters worse, your poor, neglected left hand would actually drop in temperature as much as *30 degrees* relative to your right hand.

THERE ARE MANY IMPORTANT LESSONS of the rubber hand illusion. Most obviously, it shows the influence of visual information on how you interpret touch, and how your brain changes accordingly. But just as importantly, the illusion adds to the evidence that what you *think* influences how you and your brain interpret touch. Critical to all the brain effects of the rubber hand illusion is that you must experience the hand as your own. There are ways to make this illusion fail. As stated, if the real and rubber hands are stroked *asynchronously*, the illusion and its brain effects will not occur. Similarly, if the rubber hand is positioned so that it doesn't look connected to your body, no illusion occurs, no plasticity is induced, and no disowning effects (temperature drop) ensue. In fact, subjects' ratings of the illusion's intensity—the extent to which they experience the rubber hand as their own—closely predict all of the physiological effects.

To a surprising degree, the map of your touch brain represents the *apparent,* rather than physical, aspects of touch.

SEEING YOUR BODY TOUCHED CAN GIVE YOU AN OUT-OF-BODY EXPERIENCE

THERE IS A WAY to make your whole body into a rubber hand. And the method allows you to have an out-of-body experience without dying, drugs, or psychopathology. Imagine yourself in this experiment. You sit in a chair and wear video goggles: goggles that contain a small video monitor positioned just a few inches in front of your eyes. On this monitor, you see a live image of—yourself. Specifically, you see yourself from the back, including the back of your head and torso. To achieve this, a camera is placed five feet behind you, about level with your eyes. The camera feeds a live image of your back into the video goggles.

Now comes the really weird part. An experimenter enters with two sticks and then stands to your right. He starts to slowly stroke your chest with the stick in his right hand. (While you can feel these strokes, you cannot see them, because the video goggles block your view.) With his left hand, the experimenter reaches behind you and uses the other stick to simultaneously stroke the air at a position just in front and below the camera. As you look through the video goggles, you now see your own body (from behind), but cannot see the strokes administered to your chest, because your body blocks the view. But you *do* see the upper left arm of the experimenter as it strokes the air just below the camera. In a very real way, the experimenter's *left* arm appears as though it is holding the stick that strokes your chest.

If you're like most subjects, the effect would be startling. You would experience yourself as being separate from the body you're watching. You'd strongly feel as if you were at the position of the camera behind you. Subjects typically report that they feel "outside of their body," and "sitting behind themselves." And in a rather sadistic confirmation of the effect, a visible hammer blow to the empty location just below the camera induces a measurable skin moisture response in subjects, along with an experience of panic.

Similar to the rubber hand illusion, this out-of-body effect is dependent on synchrony between the visible and felt stroking. If applied in an asynchronous manner, the effect disappears. The visible and felt stroking must be interpreted as part of the same stimulation: plausibility is again a prerequisite for the effect to work. It's as if your touch and visual systems are constantly working together to give you a sense of which body you occupy, which limbs you own—as well as where and who you are.

JOHN BRAMBLITT'S UNIQUE EXPERIENCE has provided him an exceptional sense of touch. It has also, no doubt, changed his touch brain in ways that facilitate his skill. But we now know that all brains change aspects of their structure with experience. And we know that this experience need not be as elaborate as Bramblitt's. Your brain can change to accommodate even the simplest experience. And with enough repetition, these changes can be long-lasting. Regardless of your age, your brain is ready for you to establish some new expertise and can allow you to maintain that expertise indefinitely.

Helen Keller, who was blind and deaf, touched faces to understand speech.

CHAPTER 7

Touching Speech and Feeling Rainbows

IT MIGHT SEEM AWKWARD HAVING YOUR FACE TOUCHED BY SOME-
one you've just met. But not with Rick Joy. His warmth and good humor
quickly dispel any awkwardness. And there is good reason for Joy to
touch my face. It's the only way he can understand my questions. Joy is
deaf and blind. And to communicate with people who do not know sign
language, he touches their face.

I ask Joy, who is in his late sixties and has an animated, congenial face,
how many deaf-blind individuals in the U.S. use this *Tadoma* method of
lipreading from touch. Joy has his thumb resting vertically on my chin
so that its tip is just touching my lower lip. His index finger is curled
under the bottom of my chin, and his three other fingers gently touch
my Adam's apple.

Joy answers my question by naming the individuals he knows of who
use the technique. He then concludes, "I think there are about 10 deaf-
blind people in the country who still use Tadoma."

There used to be many more. The Tadoma method was taught widely
to deaf-blind children in the early to mid-twentieth century. Helen Keller
was its most famous user. Most deaf-blind individuals use tactile signing
as their chief means of communication. Tactile signing involves placing
one's hand over the hand of a signer to feel the signer's manual gestures
and finger spelling. But to communicate with individuals who do not

know sign language, the deaf-blind have few options. And before the advent of devices that could quickly translate typed text into Braille, Tadoma was the only method.

But today, there are very few Tadoma users. There are a number of reasons for this. The availability of portable text-to-Braille translation devices makes it easier for nonsigners to communicate with the deaf-blind. Also, the causes of deaf-blindness have shifted so that it is more rare and, when it does occur, is often accompanied by cognitive disabilities. This makes the intense concentration needed to learn and maintain Tadoma skills prohibitive for many individuals. Other reasons the Tadoma skill is waning include the growing number of deaf-blind individuals receiving cochlear implants, and the choice of many deaf schools to deemphasize communication techniques other than sign language. Still, the fact that speech can be perceived through touching a person's articulators fascinates speech scientists.

I CONTINUE MY INTERVIEW with Joy as he cradles my face. I speak slowly and clearly, but resist speaking loudly. "Are some people easier to understand than others?"

"Yes. Low voices are easier than high voices. It's easier to feel the voice vibrations with low voices. Men are easier to understand than women and children." Joy's own speech can be difficult to understand; a result of not being able to hear his own voice and it being some time since he's had articulation training. His mother is present to help with the translation.

"Can certain face features make it difficult to understand someone? Does my beard make it harder for you?" I ask.

"*Thick* beards make it harder. Your beard isn't bad."

"Thanks," I say. "I was thinking of shaving it off to make it easier for you. I'm glad I didn't." We both laugh.

Rick Joy lost his hearing when he was two years old, to meningitis. The infection also affected his eyes so that he lost all vision in one eye soon after, along with much of the vision in his other. Any residual vision was gone by his adolescence. Joy was then sent to the California School for the Blind where he learned the Tadoma technique from one of the few teachers in the country.

There are many methods of teaching Tadoma to the deaf-blind. Generally, the student is instructed to place a thumb vertically across the lips of the speaker, to feel the movements of both lips. This thumb position also allows the user to feel the bursts of air released when the speaker produces *voiceless* consonants such as *p*, *t*, and *k*. The index finger is often placed on the lower cheek of the speaker to both feel the movements of the jaw and the brief, subtle cheek inflations which accompany some consonants. The remaining three fingers are placed next to the vocal chords to feel the vibrations created with vowels and *voiced* consonants such as *b*, *d*, and *g*. Experienced Tadoma users often modify this technique depending on hand size and other considerations. Rick Joy believes that he can understand just as much without placing his thumb across both lips, a position he feels can make some talkers uncomfortable.

Tadoma training with deaf-blind children typically involves a game of imitation. Assuming the children have some concept of language (either through tactile sign or Braille), they're first told they will learn a technique to understand the speech of nonsigners. (Using Tadoma to teach nonverbal deaf-blind children the *concept* of language has also been done.) The children are instructed to touch their teacher's face, using the described technique. The teacher then repeats a single syllable (*ba*) multiple times. The children's hands are then guided to their own faces with the intent of having them repeat that same syllable. The teacher might then define this syllable to the students using the students' familiar method of language (tactile finger spelling, Braille) and then repeat the imitative face touching for that syllable. This imitation procedure is then used for other syllables, and then words. Once comfortable using Tadoma to recognize words, the children attempt to recognize sentences from their teacher's face, ultimately leading to conversations. Depending on the children's age and general language skills, it can take months or years of practice to establish the skills to converse using Tadoma with novel speakers.

This imitation method of training Tadoma has the added benefit of improving the children's language production skills, by allowing the children to compare their own felt productions to those of their teacher.

■　■　■　■　■

DESPITE HIS HANDICAPS, Joy thrived as a teenager and young adult. He became an active swimmer and (tandem) cyclist and was the first deaf-blind Eagle Scout in the U.S. Upon graduating from the California School for the Blind, Joy moved to New York where he learned electronics, specializing in radio construction. He was ultimately hired by Hewlett-Packard in California where he worked for 30 years. During that time, he went back to school to earn his high-school equivalence and associate's degrees.

I ask Joy if he's had conversations with other deaf-blind Tadoma users.

He responds enthusiastically, "Yes! I talked with Tad Chapman. Tad was one of the first two people taught Tadoma. He's the Tad in *Tad*oma!" (The other person first trained in the technique was *Oma* Simpson.)

As I continue talking with Joy, his four-year-old daughter bounds into the room and sits next to me. Joy asks his daughter, "Jackie, can you get Mommy, please?" (Jackie's hearing and vision are normal.)

I ask Joy how he knew that Jackie came into the room.

"With my feet. I can feel her bouncing on the floor." Joy smiles as he stomps his own feet to show me how the floor conveys vibrations. "And I can feel the air moving when someone passes by. My boss would be surprised that I knew when he was near me. When he approached, I could feel the air move on my face, neck, and back."

Joy's wife, who is deaf but not blind, comes into the room and Joy talks to her using tactile sign. Joy gently cups his wife's hand with his own as she signs and finger spells into his palm. He responds by nodding his head, or for longer responses, using his hands to sign in front of her. The interaction seems quick, spare, and intuitive—similar to any obligatory chat between husband and wife.

Joy then proceeds to give me a tour of his electronics workshop, many of whose components he built himself. He is an avid amateur radio operator who communicates with other radio hobbyists around the world. I ask how he does this, and he picks up a custom designed box containing an open, hand-sized loudspeaker. "I read Morse code by feeling the dots and dashes as vibrations on the speaker."

He then shows me his computer, which is connected to a device that translates the screen's text into Braille. The Braille is formed in an array of raised plastic dots that changes as each new line of text is selected.

Joy's computer also has an audible text reader that uses a synthesized voice to speak the text out loud. When I ask why he would need this, he smiles and bends down under his computer desk to pull out a homemade foot rest. He turns the foot rest over to reveal a large loudspeaker that is connected to the computer and vibrates to the synthesized voice. "I feel the voice with my feet so I know where there's text on the screen. Once I find text, I highlight it and have the computer translate it to Braille."

The tour continues and Joy shows me a covered patio he designed and constructed himself. "I do a lot of work around the house. My neighbors get nervous when I work on the roof."

I'm struck not only by Joy's ambition and warmth, but by how well we are communicating. Despite having to lift Joy's hand to my face before I speak, our interactions seem natural, informative, and friendly. Certainly, I often need to repeat myself for Joy to understand me. But I'm not repeating myself more than I would for someone with a mild hearing loss.

THIS JIBES WITH THE SCIENTIFIC research on Tadoma. Practiced Tadoma users are as adept at understanding speech from touching as is a hearing person listening to speech in a noisy environment. And many of the same compensatory tricks work in both contexts. The Tadoma user is aided when a speaker talks slowly, repeats, and uses simpler, predictable words. Some types of speech movements are easier to perceive than others using Tadoma. Consonants that involve lip and jaw movements (*p*, *m*) are relatively easy to perceive. In contrast, consonants distinguished only by tongue position are more difficult to feel from the face (*k* versus *d*). However, Tadoma users are able to use the meaningful context of words and sentences to disambiguate difficult consonants and vowels.

While the skills of experienced Tadoma users such as Rick Joy are impressive, Tadoma is not a skill accessible only to the deaf-blind. Research shows that with enough training, anyone can learn to effectively understand speech from touching faces. If you have about 100 hours to dedicate to Tadoma, your identifications of consonants and vowels—as well as *sentences*—would be comparable to deaf-blind users.

But even right now, without any practice, you can perceive speech from touching faces. Research shows that touching a talker's face in the

Tadoma manner can help you better understand speech when heard against a background of noise. And if you were asked to visually lip-read, simultaneously touching the articulating face would enhance your understanding. This is true even if you've never intentionally perceived speech from lipreading or touching before. Your brain is ready to use speech information from all sources: even sources you didn't know about until today.

In fact, while you've likely never Tadoma-ed before, your brain can already use felt speech well enough so that it can *override heard speech*. Imagine being in this experiment. You stand just to the side of an experimenter who is out of view but within your reach. The experimenter brings your hand to his face so that it is in a (modified) Tadoma configuration. Your task is to feel his face utter simple syllables as you hear synchronized audio syllables emanating from a nearby loudspeaker. You're told that the syllables you touch and hear will sometimes be the same, and sometimes be different. Your job is to identify both syllables on each trial. You're also told that the presented syllables will be either *da* or *ga*— an easy distinction for Tadoma novices to feel (try it yourself). You then touch, listen to, and identify the syllables 60 (randomized) times.

The results would show that the syllables you report *hearing* are strongly influenced by the syllables you *touch*. Many of the trials would involve the experimenter mouthing one syllable, *ba* for example, while the loudspeaker simultaneously played another syllable: *da*. On these trails, you'd often report *hearing* the syllable you *touched*—*ba* in this example. This influence of touched speech over heard speech is impressive especially considering how much more experience you've had hearing versus touching speech. In fact, experience seems to have little to do with how heard speech can be influenced. While rarely experienced touched speech can influence the speech you hear, research shows the oft experienced text presentation of the syllables (seeing the written text *ba*) *cannot* influence heard speech in this way. (You'll learn why this occurs in Chapter 10.)

So despite its novelty, the speech you touch can override the speech you hear. And, as stated, the speech you touch can enhance your understanding of speech heard in a noisy environment, as well as lip-read speech. Despite not having the experience of Rick Joy, your brain is ready to integrate the speech information available from touching a face.

And the information you can get from touching faces extends well beyond speech.

You Can Recognize Faces and Their Expressions from Touch

CLOSE YOUR EYES AND use your two hands to touch your face. Feel the breadth of your forehead, the ridge of your brows, the shape of your nose, and the cut of your jaw line. Do you think that you'd now be able to recognize a plaster-casted mask of your face from touch? Research suggests you'd have little trouble choosing your mask from a set of seven. Now imagine the face of a friend. Do you think that you'd be able to pick out his or her casted mask from touch? Research suggests you would. In fact, you need very little visual familiarity with a face to recognize it from touch. Five minutes of looking at a set of photographed faces would be enough for you to recognize them through touch.

New research is showing that facial recognition is not just the purview of your visual system. You can recognize faces from simply touching them. Imagine this experiment. You are blindfolded and earplugs are placed in your ears. A strong-smelling ointment is applied just below your nose to mask any odor cues. Your hands are then guided to a real, live human face. This face is bordered by a head band so that you can't feel its hair and the eyes and mouth are closed. As you've been instructed, you explore this very brave face with both of your hands for as long as you like. Once satisfied, your hands are removed and then guided to touch three (real) comparison faces, in succession. One of these three comparison faces is the same as the original face you just touched and your job is to determine which. After your judgment, you're asked to perform this entire comparison two more times, with different, and equally brave, faces. Despite the awkwardness of this task, you'd likely be outstanding at recognizing the target face: 80 percent of the time, if you're like the research subjects.

You'd also be nearly as accurate recognizing a real face's *expression* from touch. Rick Joy commented that as he performs Tadoma on a talker, he can also easily feel the talker's emotional expression and whether it jibes with what the talker said (to determine more complex emotiolinguistic conveyances such as sarcasm). But even novice face touchers are able to recognize the expressions of happy, sad, angry, fearful, and surprise, with high accuracy. Interestingly, the ability to recognize facial expression is much greater when subjects feel a face *dynamically* producing the expressions (changing from a neutral to emotional face, often repeatedly). When feeling faces that maintain expressions statically, subjects are not quite as accurate. The fact that facial *movements* provide important information will be discussed in Chapters 8 and 9.

YOUR ABILITY TO RECOGNIZE FACES and their expressions from touch is interesting in its own right, and constitutes a hidden perceptual skill. But the scientific implications of these findings are important. They suggest that facial perception, like speech perception, is such an integral part of the human repertoire that your brain cares little from which sense face information comes. While there is no doubt that you recognize faces more easily from seeing than touching them, it's likely that your brain uses many of the same mechanisms and strategies for both.

Here's an example of how. It's well known that you are especially efficient at recognizing faces when you see them upright versus upside down. While this might not seem surprising, the full story is more interesting. Inverting facial images makes them *disproportionately* harder to recognize than inverting images of other types of objects (houses, airplanes, dogs). There seems to be something about upright faces, possibly the configuration of the features relative to each other (your eyes are this far above your mouth), that makes facial recognition especially efficient. But when faces are inverted, that configuration information is lost. Because you don't use as much configuration information to recognize other objects, their inversion is less disruptive.

This has long been understood as a *visual* facial perception effect. But recent research has shown a similar effect when you *touch* faces. Turning a face upside down makes it *much* harder to recognize from touch. In contrast, turning another object (clay teapot) upside down makes it *no* harder to recognize. Your brain uses upright information to efficiently recognize faces whether you are doing so from sight or touch. It uses the same specialized strategy for both senses.

And there is evidence that similar (or at least related) parts of your brain are used to recognize faces from vision and touch. Some of this evidence comes from research on brain injury. In one example, an individual whose injury induced an inability to visually recognize faces—an ailment called *prosopagnosia*—was asked to recognize faces from touch. This prosopagnosic found the touch task nearly impossible, performing at chance levels and substantially worse than subjects without brain damage. In fact, unlike the control subjects, the prosopagnosic was just as poor at using touch to recognize upright as upside-down faces (a similar effect occurs when prosopagnosics are asked to recognize faces from *sight*). Thus, this prosopagnosic's brain injury affected his recognition of faces from sight and touch in a way that hindered his use of upright face information for both senses. This could mean that the same specialized brain area is used for facial recognition from both sight and touch.

Recent brain imaging research has provided evidence consistent with this idea. When you look at a face, a particular part of your brain is "uniquely" activated. This small region, the *fusiform gyrus*, is located roughly between your ears. It is known to react to facial images, but not much else. (Although, as you'll learn in Chapter 8, the story is more complicated.) But recent brain imaging research shows that the fusiform gyrus is activated when you attempt to recognize faces (masks) by *touch* as well as sight. Further, activation of this area from touch occurs only when the stimuli are facial masks, not abstract, ceramic control objects. This would seem to be good evidence that you use your *fusiform gyrus* to recognize faces through vision and touch. But there is a small hitch. You actually have *two* fusiform gyri, one each on the two sides of your brain. And while *seeing* faces activates your *right* side's fusiform gyrus, *touching* faces activates your *left* (assuming that you're right-handed). This could be because the left side of your brain is generally better suited for the *sequential* exploration used when you touch versus look at something. Regardless, it's a bit premature to conclude that the same specialized brain mechanism is used for facial recognition through touch and sight.

But the overall lesson should be clear. You are surprisingly good at perceiving faces from touch. Your skills range from recognition of facial identity and expression, to the more subtle movements of speech articulation. And practice can allow you to understand speech at usable levels from touching faces. So faces can provide rich information for your touch system. This likely has as much to do with faces as it does touch. As you'll learn in Chapter 8, there is no natural stimulus that conveys as much information, and with such efficiency, as the human face. And your brain is designed to take advantage of this fact. Faces are so important to your brain, it can learn about them through any means—and any sense—available.

YOU FEEL THINGS WITHOUT TOUCHING THEM

CLOSE THIS BOOK. But before you do, remember these instructions. (If you have another book nearby, you can perform the demonstration *as* you read these instructions.) Close the book and then close your eyes. Now, gently fling the book down on some nearby surface, giving it a gentle spin and flip as you release it. The goal is to have the book land in an unpredictable position. With your eyes still closed (or, if you're using a second book, looking only at these instructions), reach down and touch the book with the tip of your index finger. Leave your finger in that spot and then put your thumb on the underside of the book just opposite from your index finger. (You may need to slide your index finger toward an edge of the book to reach your thumb underneath.) You should be touching the book with only the pads of your index finger and thumb tips—nothing else. Using just these fingers, pinch hard to get a good enough grip to lift the book. Slowly lift the book and then turn it to the side so that it hangs down from your two fingers. Maintain a strong grip on the book. Now, using only the motion of your wrist, slowly swing the book back and forth—a few inches in each direction.

Without even looking at the book, you should have a good sense of where you've gripped it, as well as its general orientation relative to your hand. You should have a sense of whether you've gripped the book along one of its side edges or along the shorter top and bottom edges. You should also have a sense of *where* along that edge you're gripping: more toward its middle or close to one of the book's corners. And assuming that you've not grabbed it precisely in the middle of that edge, you should be able to determine whether a majority of the book's bulk is near your palm or beyond your fingers. Swing the book again using your wrist and now concentrate on the book itself. You have a sense of its general size and *shape*: the relative sizes of its vertical and horizontal dimensions (even though you probably had visual sense of this before). Holding the book at two very small points gives you a sense of its size and shape, where on the book your fingers grip, and the angular position of the book relative to your fingers.

How could you feel so much about the book while touching so little of it? It's because when you wield something with your hand, you implicitly "feel" it with your tendons and muscles more than with your skin. The muscles and tendons of your wrists and arms are infused with little sensors (mechanoreceptors). These sensors respond to how your hand moves and what makes your hand move. The way you hold the book stretches and twists the sensors in ways specific to the book's size, where it's gripped and its orientation relative to where it's gripped. And this pattern of strain helps your nervous system contract your arm, wrist, and hand muscles in just the right way to hold the book effectively.

But the information gleaned by your muscle-tendon sense supports more than your ability to maintain an effective grasp. It allows you to perceive an object's properties to perform more complicated behaviors. Your muscle-tendon sense allows you to perceive the length of your toothbrush so that you can guide it to your mouth; choose the right hammer to drive in a small nail versus large spike; and feel the sweet spot of your tennis racket to most effectively return a serve. Further, these sensors inform you about the *orientation* of your toothbrush, hammer, and racket so that you can position their business ends with the correct angle. You use your muscle-tendon sense all the time; it's actually more common than the touch you use to feel the texture of something. And you implicitly gather information using this sense with little conscious awareness, leaving your eyes to attend to more distant and more important things.

Over the last 20 years, research has revealed the astonishing detail that muscle-tendon sense can provide. This research shows that simply by grasping a very small part of an object and wielding it back and forth, you can determine a great deal of information about it. For example, with your eyes closed, you could grasp the end of a wooden or metal rod and wield it for a few seconds to determine its length with impressive accuracy. If you instead grab the rod somewhere in the middle and twist your wrist from side to side, you'd be able to perceive the rod's overall length, as well as the lengths of the portions on either side of your hand. If the rod happens to have a significant bend, you'd be able to determine this fact by wielding, as well as the direction of the bend relative to your hand and arm. Remember that in all of these examples, you'd be holding only a very small portion of the rod. Yet, despite how little of it your skin touches, you'd be able to perceive properties of the rod's entire expanse. Your muscle-tendon sense allows you to feel much more than your skin can touch.

Astonishingly, your muscle-tendon sense can even tell you about the *shape* of an object without touching it. Imagine being in this experiment. You stand next to a table which is divided by a tall curtain. On the left side of the curtain sits a set of five wooden objects: a cube, hemisphere, cylinder, pyramid, and inverted cone. These objects range in height from 6 to 16 inches and have been designed to be comparable in weight and volume. Connected to the bottom of each shape is the same-sized wooden dowel, five inches long and a half inch thick.

You're positioned so that most of your body is in front of these shapes, but your right arm extends to the other side of the curtain so that it can't be seen. You are asked to open your right hand and grip the wooden dowel (five inches long and a half inch thick) that has been placed in it. You are told that this dowel is connected to one of the same shapes you see in front of you and that your job is to determine which. You must do this while holding only the dowel, but are encouraged to use your wrist to wield the object to and fro. You make your guess, and the object (and dowel) is taken from your hand. You are then given another object, which is again attached to the same-sized wooden dowel. Again, by holding only this dowel and wielding, you must determine which of the objects is connected.

Believe it or not, simply by swinging your wrist in this manner, you would be able to recognize each of the five shapes with impressive accuracy. Further, you would be accurate even if an additional smaller set of the objects were mixed into the group. This fact means you aren't simply using the apparent *weight* of the objects for your recognition. Instead, you somehow feel the shape of objects without touching them.

How does this work? When you lift something, your muscle-tendon sensors detect how the object *resists motion* in different directions (the object's moments of inertia or inertia tensor). Try wielding the book again. This time, however, use all your fingers to get a good grip, and wield the book in many different directions. You'll notice that it is more difficult to wield in some directions than others: up and down versus left to right, for example. This pattern of directional resistances depends on the book's size and shape, as well as its orientation. And it is this complex pattern of motion resistances that your muscle-tendon sensors detect. This information is used by your brain to gauge how and where your hand should effectively grip an object. It is also information you can use for more conscious purposes such as determining the length, size, and shape of what you're wielding.

You're also able to perceive more practical properties of objects with your muscle-tendon sense. Research suggests that you are superb at judging a tennis racket's (primary) "sweet spot" by simply wielding it. The sweet spot (center of percussion) on a racket or bat is defined as the location that can impart the most energy-efficient strike, while conveying the least vibration back to your hands. (Its position is determined by the object's centers of mass and rotation.) For a tennis racket, the sweet spot is somewhere within the area of the racket's netting; for a bat, it is somewhere along the barrel. What is true in both cases is that the sweet spot is not *visibly* distinct: you can't determine its precise position by looking. But it is something that you can readily perceive when holding and swinging a racket or bat. You do so using your muscle-tendon sense. In fact, research shows that by wielding a particular (unseen) tennis racket for a few seconds, novices are as accurate as experts at identifying its sweet spot. And if tennis isn't your thing, your muscle-tendon sense can also help out with hockey sticks. With simple wielding, both expert hockey players and novices can choose the best (unseen) sticks for making power versus precision shots.

You also use your muscle-tendon sense when lifting tools. Research shows that you'd be able to judge the sweet spot on a handheld bar so as to best perform *power* hammering—like you might use to drive in a spike. You'd also have no problem identifying the best position on the bar to perform *precision* hammering—as you might use on a small finishing nail. This best position for precision hammering turns out to be much closer to your hand than is the bar's sweet spot. (Precision hammering doesn't require reducing the vibrations conveyed to your hand.)

But lest you think that your muscle-tendon sense abilities are limited to your wrist, many of these perceptual skills can actually be performed with other parts of your body. You can perceive the lengths and bends of rods by wielding with your elbow or shoulder joint, rather than wrist. And even using your knee or ankle to wield a rod (attached to your foot) allows you to perceive its dimensions: useful knowledge for the next time you wear clown shoes. On a more fine-grain scale, twiddling a toothpick-thin rod between your thumb and index fingers can provide information about its length. In all these instances, your muscle-tendon sensors pick up on the same informational properties: the resistance of the rods to movement in different directions.

The important lesson is that when it comes to perceiving the things you hold, what happens *deep under* your skin is at least as important as what happens *on* your skin. Your muscle-tendon sensors are stretched and twisted in ways that tell your brain about the objects you hold. This helps your brain know how to effectively grip something, and manipulate it in useful ways. You can feel things without touching them, and this helps your touch system work more efficiently.

You Touch with More than Your Body

Madonna would be great at fly-fishing. So would Jennifer Lopez and Ann-Margret if they have any interest. According to Rick Passek, "Women with no fishing experience, that have an artistic and rhythmic sense, are definitely the best learners." Passek has been teaching fly-fishing professionally for 18 years. He's written books on the subject and has been fly-fishing most of his life around his native British Columbia. "Women typically pick up casting much quicker than men. Casting involves a fluid, almost graceful motion that seems to come less naturally to men—especially men who have experience with other styles of fishing."

In preparing the typical fly-fish cast, about 25 feet of line is first pulled out from the reel. The rod is then held up over the head and smoothly swung back so that the line trails the motion of the rod tip. Just as the full length of the line reaches its backmost position, the rod is flung smoothly forward (and more line is released). This allows the line to fly forward and out toward the target position in the water. The trick is the fluidity of the back and forward movement, as well as the timing of when to switch the rod's direction. "It's critical to feel for the small tug when the full line 'uncurls' and is extended behind you. Feeling that tug tells you it's time to bring your rod forward, and release more line," Passek says. "If you bring the rod forward before the tug, or wait too long after it, your cast will fall apart and the line won't go anywhere. It takes some folks awhile to learn how that small tug feels in the rod. I'll tell new students to first look at the line as they practice their casting. But after awhile, most folks won't need to look at the rod or line at all. I never look—for me, it's all feel."

It turns out that feeling through the rod is critical to most aspects of Passek's fly-fishing. "Up here in British Columbia, the lakes and rivers are pretty murky so we can't use our vision to see the fish. We know when we get a strike, and how to fight a hooked fish through how it feels in the rod." And the amount of detail Passek can feel is truly astounding. "I've been doing this for so long, I can usually tell the type of fish I've hooked from how it feels in the rod. I just came back from the Squamish River where there are mainly four types of trout: rainbow, cutthroat, steelhead, and bull. I can almost always tell which type of trout I have on my line by how it feels. If I start to feel little 'dit-dit-dit' vibrations in the rod, I know it's a rainbow shaking his head, trying to lose the hook. If, instead, I feel more of a 'waaw-waaw' type of tugging, I know it's a cutthroat roll-ing in the water."

Passek is also usually able to tell the gender and rough age of the fish before he brings it in. "Younger fish are just more feisty. They'll do lots of tail shaking and head wagging. On the other hand, when older fish get hooked, they'll try to dive into the weeds or run in a consistent direction. And I can tell females from males by how much of the characteristic behaviors I feel them perform: females will generally do less."

Some of the information conveyed through the rod is particularly use-ful for fighting the fish. "It's critical to know when the fish is about to jump out of the water. Flying out of the water is the best thing a fish can do to lose a hook or snap the line. If you can keep him from jumping, or at least keep the line tight when he does, you'll lessen the chances he'll get loose. I teach my students to feel when the fish is racing straight towards the top—that means he's going to jump. As soon as they feel this, they need to quickly bring the tip of the rod down towards the water. We call this 'bowing to the fish.' This keeps the line taut and down, so it's harder for the fish to lose the hook. It all comes down to timing and feeling the rod to know what the fish is doing."

■   ■   ■   ■   ■

How is Rick Passek able to feel so much detail through the rod? The fish's actions get transmitted through the line, to the rod, causing characteristic pulls and vibrations at the rod's handle. Passek is then able to use his skin and muscle-tendon sense to feel these vibrations. And his 30-odd years of experience allow him to detect subtleties in the vibrations to know the specific type of fish he's hooked, as well as what the fish is doing and is about to do. But while Passek's ability is certainly amazing, it is really just an extreme version of a skill you use all the time: a skill called *proxy touch*.

Try closing the book again. Grasp it firmly in one corner and use the opposite corner to touch some nearby surface: the arm of your chair, the covers of your bed. Close your eyes, and slowly glide the book's corner across the surface so that you have a good sense of the surface's texture. Now bring the corner of your book to a different type of surface and do the same thing. You should be able to feel this second surface as having a different texture from the first.

Two things should be apparent from this demonstration. First, you can get a good sense of a texture without directly touching it. Second, as you hold the book, your immediate experience is of the *surfaces* being explored, not of the small vibrations in the book, as such. This is true despite the fact that your skin is only in contact with the book.

You actually touch things in this way all the time. Think of other examples. By holding a spoon to stir, you can determine whether a soup is sufficiently thick or has started sticking to the pot. You can feel through a held pencil whether you're writing on gloss, standard, or cotton-weight paper. This allows your hand to apply just the right amount of pressure to effectively use your pencil. And recall a particularly impressive example from the previous chapter. Blind painter John Bramblitt is able to identify a paint's color by using his brush to feel its viscosity.

Using a probe or tool to touch something is called *proxy touch*. And whether using a knife to butter toast or a brush to groom your hair, your touch experience is of the texture the tool touches. The mechanical action occurring at the tool's tip gets transmitted through the tool to your skin, allowing you to perceive properties of the surface. Research is revealing the detail and variety of things you can perceive through proxy touch.

First, depending on the type of tool you hold, your sensitivity to a surface's texture can be comparable to that of your skin. Relevant experiments typically have subjects hold a five-inch-long plastic stylus that is about a sixteenth of an inch thick. They are asked to draw the tip of the stylus over a number of plastic gratings composed of different groove widths and spacings. (Subjects don blindfolds and listen to white noise through headphones to ensure that judgments are based on the touch input alone.) Subjects typically rate the roughness of each grating using a numeric scale. Results show that subjects are able to discern a wide range of textures using the probe. And for many textures, ratings are comparable to those made when subjects touch the gratings directly with their index finger. Texture perception using a probe is also largely independent of the speed and force used to drag the probe across the surface, a fact that also parallels direct fingertip touch.

You can use proxy touch to perceive many other things. It's long been known that blind individuals use their cane tips to feel an upcoming curb or doorway. But there is evidence that despite your lack of experience, you have similar skills. Without practice you could accurately judge the height of a movable table by tapping a handheld rod against its surface. In fact, your judgment accuracy would likely be within four inches of the actual height, even if rod length were also manipulated. Next, you would be similarly accurate in judging the *distance* of a movable wall by extending a long wooden rod out toward it. For both of these tasks, you'd also be able to simultaneously judge the lengths of the held rods themselves.

You're also able to use proxy touch to judge things in relation to your own body's dimensions. For example, you can swing a long wooden rod to feel whether two vertical barriers are separated just enough for you to walk between them. The passable width of an aperture depends, of course, on the width of your own body. And the research suggests that subjects with different widths are able to use proxy touch to perceive the specific apertures that they, themselves, can fit through.

You can also use a long rod to determine whether a gap in the floor is just too wide to step over. The point at which a gap's width becomes too great to traverse depends on your specific leg length. And the research suggests that proxy touch allows you to determine this specific point with impressive accuracy. In fact, your ability to perform this task would be largely impervious to changes in the rod's length and weight. And despite having little, if any, experience perceiving a gap in this way, your performance would be as good as that of blind individuals who have much more experience using proxy touch (with canes) in this way.

Finally, the research suggests you can use a long rod to accurately perceive whether a surface's incline is just too great for you to stand on. In fact, your accuracy using proxy touch would be comparable to that for vision.

■ ■ ■ ■ ■

RICK PASSEK'S YEARS OF EXPERIENCE provide him the proxy touch skills to be a superb fly-fisher. But the fish don't do too badly either. "We mostly do catch-and-release fishing up here. [Passek, who doesn't like the taste of fish, is exclusively a catch-and-release fisherman.] There are older fish in the rivers that have been caught fifteen, maybe twenty times. They know lots of tricks. One favorite trick is to swim straight to the boat as soon as they get hooked. They'll go right to the boat's anchor line and then start swimming around it. Doing this wraps the fishing line up so you can't pull them in."

It could very well be that in this scenario, the fish itself is using a kind of proxy touch. By feeling the directionally specific tension in the line, the fish could have a sense of from where the line emanates. From a hooked fish's perspective, the direction of least line tension is toward the fishing rod. By following this direction, the fish can then approach the boat and anchor line to perform its clever tangling maneuver.

Speculative as it is to infer a fish's perceptual experience and strategies, there is good evidence that many animals use a type of proxy touch all the time. In fact, proxy touch is a built-in feature for many animals. The best known example is whiskers. Whiskers help small animals navigate in dark, cramped quarters. As a rat sticks his nose into a small dark hole, he uses his whiskers to get a sense of whether his body will fit through. Whiskers only have touch receptors at their base (follicles), and these receptors react whenever a whisker touches something. It's as if whiskers are a set of thin held rods, like the ones used in experiments to determine distance and width. Rats are expert as using this system, and have been shown to navigate mazes using only their whiskers. They also use their whiskers to discriminate textures and to determine their head's position relative to the surface of the water in which they're swimming. Clipping the whiskers of rats and cats has been shown to disrupt their locomotory skills in the dark.

It's likely that other animals can use their appendages as proxy touch systems. While claws, horns, beaks, hair, feathers, and antennae all have other primary uses, each could provide animals with an on-board proxy touch function. But before you get jealous, it could very well be that your fingernails can serve the same purpose. Perhaps you've noticed that it feels different filing your nails with the rough versus smoother side of an emery board. And carpenters will sometimes use the tip of their nail to test whether a small ridge on a wooden surface has been sufficiently sanded down. Research shows that humans can detect very fine ridges using only their nail tips. Because your nail only contacts touch receptors where it overlaps with your finger's skin, experiencing textures with its tip constitutes proxy touch.

Finally, there are your teeth. While you probably don't think of your teeth as sensing much more than pain, there's a degree to which they provide proxy touch of food to sense textures. Part of the pleasure of slowly chewing on taffy, or crunching a crisp carrot, comes from the proxy touch experience provided by your teeth. There's even evidence that if you hold a small plastic probe between your teeth and glide its tip across a surface, you can discern the surface's texture nearly as well as if you held the probe with your fingers!

Certainly, proxy touch is far from the primary purpose of your nails and teeth. But the fact that they can be used for this purpose evidences your touch system's general strategy to use proxy touch when it's available.

But even when you do use an external tool for purposes of proxy touch, there is a degree to which that tool becomes *part of you*. Recall from the previous chapter how both you and your brain become convinced that a rubber hand is your own. The specific characteristics of the hand—its length, position—get integrated into both your perceptual experience and how your touch brain represents your body. And as mentioned, short-term use of a tool—tweezers, tongs, pliers—can similarly induce systematic changes in your touch brain.

You Can Use Vibrations like a Spider

HERE'S ONE MORE PARTICULARLY colorful example of your proxy touch skills: you can use vibrations like a spider does. Many spiders use their webs as proxy touch tools. By detecting vibrations through its strands, spiders know when a bug has been snared in the web. In fact, some spiders construct their webs to specifically amplify the vibrations created when their most nutritional prey (typically flies) is snared. But spiders are able to use web vibrations as more than a snare alarm. Patterns of vibrations can tell the spider *where* in the web the prey has been caught. When an insect is ensnared, its mass vibrates the web in a manner specific to its location. This vibratory code is detected by the spider who can then more quickly apprehend its prey and lessen the chance of escape.

And you can do the same type of thing. You can use a strand's vibrations to determine the location of an attached object. Imagine being in this experiment. You are positioned next to a tall barrier that occludes most of a nine-foot-long, horizontally suspended rope. This quarter-inch-thick rope is pulled taut and is located about waist high. You're asked to use your right hand to grip the rope behind the barrier. From your position you can see neither the rope nor your hand. With your hand on the rope you start to feel the rope gently shake up and down. You're told that somewhere along the rope in front of your hand, a small, two-and-a-half-pound metal disk has been attached. You're also told that an experimenter is gently moving this disk up and down to subtly jiggle the rope. Your job is to feel how the rope jiggles and, from this, guess how far down the rope the disk is located. After your guess, you briefly remove your hand while the disk is shifted to a new position. You then grab the rope again, feel it shake, and guess the disk's new position. (You make your distance judgments by using your left hand to adjust a visible pointer which hangs from a pulley system.)

Despite the oddity of the task, and even without practice, you would be impressively accurate with your judgments. Over multiple disk positions ranging from one to five feet, you would usually be accurate within five *inches* of the disk's actual position. And your accuracy would be maintained regardless of how the rope jiggled (side to side, up and down), whether you felt it jiggle just twice, or if the amount of weight of the attached disk was varied from trial to trial. Moreover, you could accurately judge the disk's position even if *you* were the one jiggling the rope with your right hand. This means that the jiggling does not need to initiate at the target for you to use your spidery sense. As long as the rope jiggles, you can tell where the disk is.

How do you do it? It turns out that to determine the disk's position, your hand detects a complex pattern in how the rope jiggles. This pattern can be roughly characterized as a relation between how quickly a wave passes down the rope (and across your hand) and how quickly the rope returns to its starting position. While as a subject, you might describe this as something like the rope's "tension," it is more complicated. The pattern your hand detects is actually based on both the distance of the attached object and the actual tension (horizontal force) in the segment of the rope you hold. In fact, while your accuracy would not be influenced by the initiation point, direction, or number of the jiggles, it *would* be influenced by the actual tension in the rope's segment that you hold.

Of course, these details are not your conscious brain's concern. Nor are they the spider's. But spiders have become expert at using the vibratory patterns in their webs to determine precise prey location. They are even known to intentionally modulate the tension in their web to suit their nutritional needs. When a spider is particularly hungry, she will increase her web's tension to better detect smaller insects—prey she would ignore when sated.

Whether you're a spider locating snared prey, a subject feeling the location of an object attached to a rope, Rick Passek knowing he's hooked a rainbow, or Rick Joy recognizing his daughter's presence through vibrations on the floor, the principle is the same. The vibrations conveyed through a semiflexible medium—web, rope, fishing rod, or floor—allows your proxy touch system to feel things from a distance.

PART V

Seeing

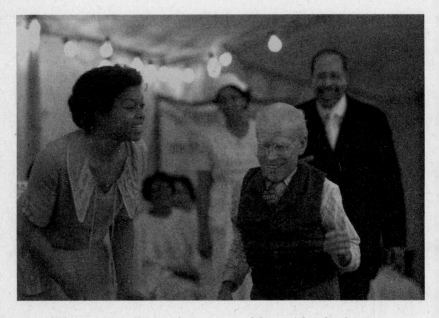

The Benjamin Button character is widely considered to have the first computer-animated face that seems convincingly real.

CHAPTER 8

Facing the Uncanny Valley

STEVE PREEG DIDN'T FEEL CONFIDENT ABOUT HIS NEW PROJECT. "This is the first project I've worked on that has a really good chance of complete failure," he told his brother. When his brother asked why, then, he had decided to take it on, Preeg answered, "Because if it *does* work, I think there may be an Academy Award in it for us." And it did work—and Preeg did win an Academy Award.

But it was a long and precarious road to Oscar night. Preeg, an animation supervisor, and the crew he was part of at Digital Domain, had been tasked with something that had never been successfully done: to computer animate a human face that truly looked real. But not only did the face need to seem human, it needed to convincingly talk and emote for 52 minutes of screen time. This animated face—and head—also needed to be superimposed on a real, filmed body that was constantly in motion. And not just any face could be animated: the face needed to look just like Brad Pitt—in old age.

When Digital Domain won the contract to create the digital effects for *The Curious Case of Benjamin Button*, Preeg was dubious about animating a convincingly naturalistic character. In fact, the entire industry was dubious. Preeg, who is in his midthirties and has a bespectacled, friendly face, remembers, "We had a really difficult time getting people to work on the project, from both inside and outside the company." And understandably so. Attempts to animate a compelling human face had been disappointing. In fact, Preeg had been involved with some of these attempts. Just two years earlier, he worked with Digital Domain on a commercial for Orville Redenbacher Popcorn, featuring the long-deceased Orville himself. Their attempt to re-animate Redenbacher's smiling, chatty face produced a character that, while having a clear resemblance, struck reviewers as "creepy," "gruesome," and the "world's first pitch-zombie." Eight years earlier Preeg had worked on a fully animated film, *Final Fantasy: The Spirits Within*. While more stylized, the pseudo-photorealism of the characters' faces also struck reviewers as "mechanical" and possessing "a coldness in the eyes."

At the same time, Steve Preeg and a crew in New Zealand had tremendous success animating *non*human characters in film. These characters included King Kong, in the 2005 eponymous film, and the creature Gollum, from *Lord of the Rings: The Return of the King*. The character animation for those films was considered fully compelling and realistic—or as realistic as a giant gorilla and reptilian hobbit could possibly seem.

So why did Preeg and his colleagues have such difficulty animating *human* faces? In fact, the problem is widespread. Attempts by other companies to animate realistic human faces had fallen to similar, and worse, criticisms. Characters in films including *Beowulf* and *Resident Evil* had all been described as being eerie and zombielike. Perhaps most conspicuous of the recent attempts was 2004's *The Polar Express*—a children's Christmas film starring a series of animated Tom Hankses. It used a state-of-the-art motion capture technology to apply Hanks's facial expressions and movements to the multiple animated characters he voiced. Still, reviewers found the film's faces "unnerving," "creepily unlifelike," and possessing a "chilly, zombie-like aura."

There seems to be something about quasi-realistic human faces that makes them eerie. This doesn't seem to be a problem for nonrealistic, caricaturized faces. Cartoon faces from Popeye to Homer Simpson have long seemed emotive and engaging, without being creepy. It's when animated faces move closer to realism that they seem off-putting.

This is, in fact, a well-known phenomenon, called the *uncanny valley*. This concept was originally discussed by robot designer Masahiro Mori. It describes how we react to seeing humanoid forms—robots, animated characters—that posses *nearly* human characteristics. If a robot,

animated character, or even stuffed animal has a few human characteristics, it can seem familiar and acceptable. If it takes on a few more
human characteristics, its sense of familiarity can continue to climb. But
when a robot or animated character's likeness gets very close to that of an
actual human, your acceptance and sense of familiarity with its appearance drops precipitously. Your acceptance of it falls into a "valley" where
its appearance seems "uncanny." Once in the valley, your reactions to
its appearance are similar to when you see a zombie or corpse—it's just
human enough for you to notice that some human characteristics are
missing or distorted. And your visceral reaction to the zombielike appearance of near-realistic humans could be a by-product of evolutionary pressures to avoid corpses, and their possible diseases.

Consider your own impressions of the famous Star Wars robots—
R2-D2 and C-3PO (the short roundish one, and the taller, bronze one).
While both are appealing, your sense of their familiarity is different—
C-3PO seems more familiar because of his more humanoid appearance.
To heighten his familiarity, you could put clothes or a wig on C-3PO—
changes that might make him even more endearing. But now imagine
putting a flesh-colored rubber face mask on C-3PO. He would quickly
go from endearing to creepy.

This is exactly why near-realistic animated faces in film have been
so unsettling. By attempting to make the characters as human as possible, animators pushed them into a range of realism in which their non-
human characteristics were eerily conspicuous. And this explains why
Steve Preeg, and the industry in general, felt such trepidation taking on
Benjamin Button. As Preeg aptly puts it: "Yea though I walk through the
uncanny valley of death . . ."

Why does the uncanny valley effect occur for *human faces* in particular? Recent research on the uncanny valley is consistent with the
perceptual uniqueness of watching human forms. There's evidence, for
example, that your brain reacts much differently when watching the
same action performed by a humanoid versus mechanical figure. Watching a humanoid robot perform an assembly-line task engages a part of
your brain thought to internally mirror the actions of another person.
These *mirror neuron* sites will be discussed in detail in the next chapter,
but suffice it to say that they are considered to be active whenever you
attentively watch, or hear, another person perform an action. Interest-

ingly, these sites do *not* seem to be activated when you watch a nonhumanoid, mechanical robot perform the exact same task. Similarly, mirror neuron sites engage when you watch an animation of a near-human figure walking, but not a stick figure walking. It seems that a degree of human realism is important for your brain to internally mirror the actions of others. Perhaps this reflexive mirroring is involved in making face perception that much more sensitive to distortions in a facial image. If so, the uncanny valley experience may be a by-product of your brain's disposition to mirror human actions.

Other research relating to the uncanny valley has examined our tolerance of specific facial distortions. It turns out that the more real a computer-animated face is rendered—using, for example, *photorealistic* skin—the less tolerant you are of small deviations in facial shape, as well as feature size and location. This fact seems to be particularly true of deviations in the size, position, and details of the eyes—reportedly one of the eeriest features of *The Polar Express* and *Final Fantasy* characters.

On the other hand, you are much more tolerant of deviations in facial shape and features if the same face is rendered with less realistic skin. Similar effects have been found for our tolerance of a character's movements: the more photorealistic the character, the less tolerant we become of less natural movements. In fact, a general all-or-none principle is emerging from this research: if any characteristic of a face is rendered in a convincingly realistic way, *all* other characteristics must be realistic for the face to not seem eerie.

And this is something Preeg and his team knew all too well going into the *Benjamin Button* project. They knew that if the Benjamin character was to be the first to traverse the uncanny valley, *every* aspect of his face—its shape, skin, features, movement—needed to look perfect. Preeg knew that this would be especially true of the eyes.

"We had originally been studying glass eyes to help us render Benjamin's eye movements. And then one of our animators took a close-up, high-definition video of his own eye moving. After closely looking at it, we noticed something interesting. When a real eye moves, the thin film of moisture covering the eye builds up more on the side toward which the eye is moving. Moving your eyes to the right produces more moisture on the right side of your irises than the left. This difference in the amount of moisture creates different glimmers on the two sides of your eyes—and this difference changes whenever your eyes move. It turns out that these changing glimmers give the eyes a more natural, soulful look. If the eyes shift to the right, but then don't glimmer a bit more on the right than left, they just don't look as real. So, we added this detail to every animated shot."

But as the all-or-none property of the uncanny valley would predict, enhancing the realism of one feature made other less realistic characteristics more conspicuous. "It was a constant process of fixing one problem and then noticing something else. Someone would notice that the lighting on Benjamin's face didn't quite match the background. We'd go back, fix this, and then notice that the skin didn't look quite right when he turned his head. The uncanny valley is an evil, evil place."

Attention to this level of detail made it difficult for Preeg and his colleagues to objectively evaluate their progress. This forced them to have frequent clip screenings to get feedback. "We would have meetings to look at the preliminary clips. *Everybody* would be asked for their input on realism: from the effects supervisors to the production coordinators [staff assistants]. Our philosophy was that everyone is an expert when it comes to faces. In fact the production coordinators probably have more expertise watching faces than I do. They're social people, I'm not. After work, they go out and look at people talk, I just go home."

And Preeg's philosophy is correct. We *are* all facial experts and attentive to the subtlest aspects of this rich stimulus.

YOU ARE A FACE-PERCEIVING EXPERT

NOTHING IN YOUR PERCEPTUAL world communicates so much information so quickly as a human face. From a face, you can rapidly determine an individual's identity, gender, emotional state, intentions, genetic health, reproductive potential, and even linguistic message (through lipreading). Faces also convey more subtle personality characteristics, as you'll soon learn. It's unsurprising, then, that your brain is supremely attuned to faces.

This fact makes you extraordinarily sensitive to subtle distortions in faces, especially faces that are familiar. Just ask Chairman Mao. For Mainland Chinese, Mao Tse Tung's face is iconic: his portrait is posted in schools, offices, and billboards. As a result, if Chinese subjects are shown an image of Mao in which the distance between his eyes is shifted just *one tenth of an inch* (1 percent of the full width of Mao's face in the photo) they will notice that something is wrong. Their ability to detect such a miniscule distortion based on their *memory* of Mao's face is astounding. But in the spirit of the party serving the people, the effect is not limited to iconic faces. Recent research suggests that you can also detect this subtle eye distance change when looking at a friend's photograph, or your own.

And you detect the presence of facial images exceptionally fast: nearly twice as fast as nonfacial images. If the face is expressing an emotion, you detect it even faster. It's also virtually impossible for you to ignore the presence of a face in your field of view. Even when performing a task as automatic as reading, faces can have a disruptive effect. If you were asked to quickly read the name of a profession—"musician," for example—your reading speed would be slowed if you simultaneously saw a face of a famous person from another profession—Hillary Clinton, for example.

Faces are such an important part of your world that your brain is ready to see them everywhere. This is why you see faces in objects that aren't faces. Clouds, rocks, mountainsides, and celestial bodies can all seem to contain faces. Facial images of the divine kind have been seen in tree bark, wall stains, and grilled cheese sandwiches. A recent study shows that most of us can easily see facial images in the front of cars: headlights for eyes, the radiator grill for a mouth. In fact, when asked to apply "personality characteristics" to car fronts, we show astonishing consistency. Cars that have a wide frontal stance with narrow headlights and windshields are judged as looking masculine, confident, mature, and maybe a little angry. Narrower car fronts, with taller windshields and rounded headlights look more childlike, submissive, happy, and agreeable. It seems that when in a dark alley, we'd rather meet a Mini Cooper than a Lamborghini.

You even recognize faces unconsciously. Imagine being in this experiment. You sit in front of a computer monitor on which, you are told, pairs of faces will be shown. One of these faces will be of a famous person, while the other unfamiliar. Your job is to press one of two buttons indicating whether the famous face is on the left or right. It sounds easy enough. But then the first pair of faces is presented and you see . . . nothing. You may notice a quick flash on the screen, but certainly no faces. In fact, the face pairs were presented for less than *one-fiftieth* of a second: well under the time necessary to consciously see something. Still, you're asked to respond, so you make what feels like a wild guess and indicate that the famous face was on the right.

After 40 trials like this, you'd be convinced you were guessing every time. You'd assume that your response performance was at chance: regardless of on which side the famous face was shown, you'd be equally likely to choose left as right. In fact, you'd be right—your overall performance would be at chance at guessing the famous face.

But looking more closely at your responses would reveal something fascinating. Your "random" choice of left or right would actually depend on your emotional response to the famous face. It turns out that half of the famous faces were of likable individuals—John Kennedy, Mick Jagger. The other half were of very unlikable individuals—Adolf Hitler, Saddam Hussein. If a presented face pair contained a likable famous person, you would typically choose that face as being famous. If, on the other hand, the pair contained an *unlikable* famous face, you would typically choose the *unfamiliar* face as famous. It seems your unconscious mind *is* able to recognize famous faces and draws your responses toward the likable ones but away from the unlikable ones.

Unconscious face perception has also been observed for individuals who can't consciously recognize faces at all. Prosopagnosia is a brain-related deficit that renders an individual unable to recognize the faces of even family members and themselves. Made famous in Oliver Sack's seminal book *The Man Who Mistook His Wife for a Hat*, prosopagnosics must rely on other cues such as voice, body size, and clothes to recognize people. While prosopagnosia was long considered to result only from brain injury (due to stroke or a traumatic head injury), there is growing evidence that it can also have a congenital basis.

Regardless of its cause, research has emerged showing that a majority of prosopagnosics *can* recognize faces at an unconscious level. Despite having no conscious recognition, prosopagnosics show greater skin moisture responses to familiar faces. They are also able to more quickly state the profession of a famous individual whose name they've read, if that name is accompanied with a face of someone of the same versus different profession (similar to the Hillary Clinton example described above). Faces are so potent, they seep into the unconscious of individuals whose brains won't even allow them to consciously recognize their own face.

■　■　■　■　■

SOME OF YOUR SKILLS with faces were probably present from birth. An hour after you were born, you preferred to look at facial drawings whose features were in the correct, versus jumbled, configuration. By the time you were four days old, you could recognize your mother's face from a video. And when you were two weeks old you were able to subtly imitate many of your parents' facial expressions. These skills are especially impressive given the very fuzzy images your infant visual system provided. But your current facial expertise didn't develop until you were an adolescent. At that time, you were able to use all of the specialized facial information you use today. And it is your ability to use this information that makes your face perception skills so sophisticated and efficient.

What is this specialized facial information? Recall from the previous chapter what happens when you see—or touch—an *upside-down* face: it becomes disproportionately harder to recognize. Inverting nonfacial images does not have this same disruptive effect. Why is this true? It turns out that to recognize a face, you not only look at its features, but you also look at how those features are configured relative to each other. Imagine looking at your best friend's face. Your friend's eyebrows are located just so far above the eyes, which are just so far above the nose, which is just so far above the mouth and chin. This *configural information* as well as facial features help you to recognize your friend. In fact, you would be relatively good at recognizing your friend's face from a photograph that was made blurry—a distortion that removes much of the featural detail of a face, but maintains the configural information between the features.

But when you see your friend's face upside down (in a photo or when hanging from monkey bars), this configural information is no longer available: the eyes are now *below* the nose which is below the mouth. This makes the face much harder to recognize. And blurring an inverted

facial photo renders it nearly impossible to recognize: neither the configural nor featural information is available.

But you lose more than vertical configuration information with an
inverted face. You become less sensitive to other details including the
horizontal configuration between features. For example, your skills at
recognizing miniscule changes in the distance between the eyes of your
own, or an iconic, photograph—the Mao effect—are eliminated when
the face is inverted. You also become less sensitive to distortions of
the individual features themselves with inverted faces. Imagine how
gruesome a face would look if its eyes or mouth were flipped upside
down in an upright photograph. If, however, you saw this entire gruesome image in an inverted orientation, you would barely notice these
dramatic distortions.

So an upright face not only provides information about the vertical
configuration of a person's features, it actually enhances your sensitivity
to other facial details. Consider this. Your memory for a facial feature—a
nose, say—is better when seen in the context of a face than when seen
by itself. This is true despite the fact that a face includes a substantial amount of distracting information—information that should make
it *more* difficult to recognize that feature than when seen on its own.
But just the opposite occurs: rather than being distracting, the upright
face context *enhances* your memory for that feature. If you see the face
inverted, however, this enhancement completely disappears.

These facts have led many researchers to conclude that faces are perceived in a special way, using special strategies and brain mechanisms.
Other researchers argue that faces themselves are not special, but that
visual expertise with a class of images *is*. We are all experts with faces.
But some of us are also experts at perceiving other classes of images. Dog
breeders are experts at looking at the subtle distinctions that distinguish
dogs of the same breed and family. It turns out that when these experts
look at dog images, they show the same visual effects (upright image
enhancement) and brain responses the rest of us display when looking at faces. The same is true of expert car designers as they look at car
images. Even naïve subjects intensively trained to visually discriminate
a set of abstract objects show the perceptual and neuropsychological
responses typically associated with faces. While there is no consensus

on whether face perception is truly specialized, there is clear consensus on the potency of faces as perceptual stimuli.

You Recognize Faces by the Way They Move

Besides Steve Preeg and colleagues' supreme attention to detail, Benjamin Button's facial realism is a result of new technologies. One of these technologies is a new motion capture system called Contour Reality Capture invented by the San Francisco company, MOVA. This technique is much more sophisticated than the one used to capture Tom Hanks's face in *The Polar Express*. It allowed Preeg's crew to capture Brad Pitt's idiosyncratic *style* of facial motion, and then animate Benjamin's face using this style. They essentially created a Brad Pitt face puppet: a face they could manipulate in any way they needed.

For this purpose, Pitt's facial motions were digitally analyzed as he performed a large set of standardized dynamic expressions. "Brad spent six months practicing about two-hundred facial expressions—learning to control his face in very precise ways," Preeg recalls. "He then came in and MOVA applied a luminous speckled paint to his face and filmed him making the expressions with 28 cameras. This allowed us to very carefully track the movements of thousands of points on his face. The tracked movements were then used to build an animation program. This allowed our animators to move the Benjamin character's face in the same way Brad would move his face. When the animators lifted an eyebrow, they would be lifting it the way Brad lifts his eyebrow. In fact we noticed that, like all of us, one side of Brad's face moves a bit more than the other. Retaining these sort of imperfections helped with the realism. And using Brad's specific way of moving helped the character seem like Brad."

Preeg's impressions are consistent with recent perceptual research. It is now known that one way you identify faces is by watching how they move. Imagine yourself in this experiment—conducted in my own laboratory. You sit down at a video monitor and are told you'll be seeing short, silent video clips of your friends' faces. Your job is to simply watch the videos and identify your friends—seven of them to be exact. You sit back in your chair with confidence, thinking that now *this* is a task you can handle. The first video clip begins and you see . . . dots. Specifically, you see about 100 small white dots randomly dispersed against a black background. Somehow, you're supposed to see a friend's face in this random pattern.

Then about 30 of the dots toward the center of the screen start to move. You're told that these moving dots were actually placed on the faces of your friends when they were videotaped. These fluorescent dots were randomly glued to your friends' cheeks, nose, forehead, as well as their lips, teeth, and tongue tip. Your friends were asked to utter a simple sentence and move their faces as they were taped. By using fluorescent lighting during taping and adjusting the video contrast, facial motion was rendered as nothing but white dots moving against a black background. (Additional nonmoving dots were added around each face so that its shape and size couldn't be determined without movement.) This technique extracts the facial features and configurations you usually use to recognize faces. It essentially retains only a face's *movement* information. So now it's your job to recognize your friends by the way their faces move.

Despite the strangeness and reduced nature of the videos, you would be able to recognize most of your friends' faces. If, on the other hand, you saw these images *not* moving (as when the video had been paused), you would be completely unable to recognize your friends, or even see a face in the dot patterns. This confirms that these videos truly contain no usable face structure (feature, configuration, shape) information. It shows that you can identify friends by simply recognizing the idiosyncratic way they move their face.

When we talk, smile, and yawn, we all have subtly distinct ways of moving our face. And research over the last 10 years has revealed that you can use this idiosyncratic movement information to help identify people. Using a variety of techniques, research has shown that especially when facial images are degraded (blurring, color distorting, inverting), seeing movement helps you recognize both acquaintances and famous faces. Seeing an unfamiliar face move can also help you remember it more easily. Four-month-old infants can discriminate faces based solely on how they move. And facial motion can also convey information about the age, gender, expression, and even attractiveness of an unfamiliar face.

In a demonstration particularly relevant to the uncanny valley, one person's facial movements can be applied to another person's facial structure. Typically, these experiments involve capturing a real person's facial motion and then applying that motion to an animated face. The technique provides researchers a way to test the relative importance of facial structure (features, configuration, shape) and movement information. Imagine Brad Pitt's facial motion captured and then applied to the computer-generated face of Tom Hanks. As you watched this face move, you would probably still identify it as Tom Hanks much of the time. But you would identify it less often as Hanks than if the face moved with Hanks's *own* characteristic motion. Imagine now that a morphed face is created from the two actors' faces. This face would have a shape, features, and configuration that looked somewhere between those of Hanks and Pitt. Your identification of this face would depend strongly on whether it moved with Hanks's or Pitt's captured motion. These effects show that facial motion can influence recognition of a face that has *not* been degraded by blurring, inversion, or white dots. And it shows that idiosyncratic facial motion is potent enough to sometimes override information from a face's features and structure.

■　■　■　■　■

STEVE PREEG'S ATTENTION TO DETAIL and Digital Domain's invention and adoption of the newest technologies paid off. Benjamin Button is considered by many experts to be the first animated character to cross the uncanny valley. He looks completely human, natural, and not the least bit eerie. And he moves, talks, and emotes exactly like an older Brad Pitt. Even other animators have stated they were unaware that the real Pitt doesn't appear until nearly an hour into the film. And for their success, Preeg and his colleagues at Digital Domain won the 2008 Academy Award for best visual effects.

Still, Preeg has trouble not being critical of his work. "I still notice things that aren't quite right with Benjamin. I cringe whenever his eyes blink. When we blink our eyes, the lower lids shift inward toward the nose about 50 percent. This is something we didn't learn until after we were done." When I ask when he realized that Benjamin had crossed the uncanny valley, he laughs and says, "During the postaward interview they asked when we knew we had accomplished something special. I held up my Oscar and said, 'Pretty much right now.'"

You Use Your Own Face to Perceive the Expressions of Others

ARGUABLY THE MOST IMPORTANT information you get from a face is what its owner feels: whether someone intends you harm or affection, needs consoling or privacy. Of course, your reactions are also communicated in your own expressions. It's as if our faces are involved in an intimate dance with one another: a dance out of which can emerge empathy, antipathy, seduction, or a struggle for dominance.

And research is now showing that much of this dance can proceed without your awareness. Your brain detects facial expressions that you don't consciously notice. And these subliminal expressions can have a profound effect on you. This research typically uses extremely brief presentations of facial photographs—presentations too quick to consciously detect, let alone identify, as faces. These subliminal faces are then immediately followed by some other image presented at a much longer, noticeable duration. Still, the subliminal faces have an influence. If, for example, you were presented with a subliminal smiling face, you would likely have a more positive opinion of a subsequently presented abstract design (a Chinese ideograph) than if the subliminal face were scowling. Interestingly, if either type of face were presented for a period long enough for you to consciously notice, it would *not* affect your aesthetic judgments. At least in this context, a subliminal expressive face has a stronger influence on you than does a face you are consciously aware of seeing.

Subliminal faces can also subtly influence your mood. If you were presented subliminal faces expressing anger, fear, or disgust repeatedly for some time, you would later rate your mood as more negative than if the faces had neutral expressions. Subliminal faces can even influence your (hypothetical) behavior. Imagine this: You are walking down the street at night and notice a suspicious person walking toward you. How likely would you be to cross the street to avoid the person? Pretty likely—if, in fact, you were first presented a subliminal face posing a fearful expression. But now imagine this scenario: You're in a foreign country and the local cuisine contains fried grasshoppers. Would you partake? Probably not—if you were first presented a subliminal disgusted face. These are two of the hypothetical scenarios posed to subjects after seeing subliminal faces. Subjects' reports of their likely behaviors were strongly influenced by the expressions of faces they never consciously saw.

Finally, subliminal facial expressions can actually influence . . . your *face*. It's been known for some time that consciously seeing a facial expression can induce subtle, imperceptible reactions in your facial muscles. When you see a smiling face, the muscles involved in producing your own smile (including zygomatic major) show subtle increases in activity. This increase is too subtle for you to feel, but can be measured with small electrodes placed on your lips.

But more recent research shows that you don't even need to consciously see a face for this reaction to occur. So while *you* don't notice a subliminal smiling face, your lips do. Subliminal smiles induce subtle but measurable increases in the electrical activity of your smile muscles. If, instead, the subliminal face were angry, your smiling muscles would show no increase in activity. Instead, the muscles that knit your eyebrows during a frown (corrugator supercilii) would show the increase. Your face subtly mimics facial expressions you don't even notice. Your parents and teachers were right: A smile is contagious.

Why does your face covertly imitate expressions? Certainly, your *overt* facial reactions to other people's expressions help convey empathy and engagement. Whether participating in a mutual smile during a greeting, or a sympathetic frown when hearing sad news, overtly reciprocating an expression helps solidify our relationships. Might your subtle, covert facial reactions simply be a by-product of overt sympathetic expressions? Possibly. But they may have a deeper basis. After all, you covertly imitate whether or not you're consciously aware of seeing an expression. And as research has shown, you do so even if explicitly instructed to *not* move your face. There is a persistence to your covert facial reactions suggesting that they have a more important purpose.

In fact, covert facial reactions may actually help you *recognize* expressions. Research shows that if your covert reactions are inhibited by either high-tech means (transcranial magnetic stimulation) or low-tech means (firmly biting down on a stick), your ability to recognize expressions would be changed. Inhibiting activity in your smiling muscles would make it more difficult for you to recognize that a face was subtly smiling. Inhibiting these muscles, however, would have no effect on your recognition of sad or fearful expressions. So not only do your covert expressions help you recognize others' expressions, they do so in a muscle-dependent manner.

Of course, covert facial reactions aren't *necessary* for you to recognize expressions. Neither high- nor low-tech methods to inhibit muscular reactions serve to completely eliminate expression recognition. And you're still pretty good at recognizing expressions when your face is engaged in other activities like chewing or talking. Yet, the effects of covert facial reactions are surprising and, possibly, profound. They suggest that you use your own face to perceive what other faces are doing. The general notion that your motor system is involved in perception is the topic of the entire next chapter. But it suffices to say that the empathetic reactions you have to facial expressions do more than help facilitate your social interactions. They very well may help you recognize expressions. And they may be involved in something even deeper.

YOUR EXPRESSIONS INFLUENCE YOUR EMOTIONS

THERE IS GROWING EVIDENCE that your covert facial reactions interact with your own emotions. For example, covert reactions are influenced by emotional context. If you were first put into a scared mood by watching a horror film, your covert reactions to seeing a fearful face would be amplified. Also, covert reactions can be induced by emotional stimuli other than facial expressions. Hearing happy or fearful voices, or seeing photographs of happy or fearful body postures, all can induce covert facial reactions. It seems your covert reactions empathize with someone else's emotional state regardless of how it's conveyed.

There's even evidence that your covert reactions can play a role in *determining* your own emotions. Imagine being in this experiment. You're told that you are participating in a research project that will help establish alternative tool handling techniques for physically impaired individuals. For your particular subject group, the tests involve holding and using a pen in ways that don't involve your writing hand. You're first asked to hold the pen with your protruded lips so that its point extends out away from your mouth—like a cigarette. While holding the pen in this way, you are asked to: (1) perform a connect-the-dots task; (2) underline the vowel letters in a random array of printed consonants and vowels; and (3) read some cartoons (*Far Side*) and underline a number on a 0–9 scale to indicate how amused you are by each. You're then asked to take the pen from your lips, and gently hold it between your upper and lower teeth so that it again protrudes with the point out. You're also told to make sure that your lips don't touch the pen. Holding the pen this way, you perform the same three tasks (but with a new connect-the-dot pattern, letter array, and cartoon set). Finally, you're asked to perform the three tasks with the pen in your nondominant hand.

How would you do holding the pen in your mouth? It doesn't matter. What matters is how you would rate the cartoons. If you're like most subjects, you would indicate that you were most amused by the cartoons when the pen was between your teeth. You'd be moderately amused when the pen was in your nondominant hand. And you'd be least amused when the pen was between your lips. Why? The answer could lie with your covert facial actions. When you were told to gently hold the pen between your teeth and not touch it with your lips, your mouth was serendipitously put into a smile. This may have subtly boosted your emotional state just enough to induce a more amused reaction to the cartoons. When, instead, you held the pen between your lips, your mouth was put into a frown. This could have subtly depressed your emotional state so that you rated yourself as less amused. The pen in your hand served as a control condition to ensure that it wasn't the difficulty of the task that determined your amusement ratings (writing with your nondominant hand is much easier than writing with your mouth).

Importantly, the subjects in this experiment were not aware of its purpose: The physical impairment cover story was completely effective. This means that subjects were not intentionally, say, raising their amusement ratings when they realized they were actually smiling. In fact, the subjects never realized they were smiling at all. Instead, the results suggest that our own facial expressions can play a role in our emotional states.

It should be noted that the effects of induced expression on emotion are subtle. It's unlikely that biting a pen can pull you out of a truly bad mood. But these effects do seem real. Similar sneaky methods to induce expressions can influence your ability to read and remember emotional words and can induce physiological changes associated with different types of emotions. Also, surreptitiously *inhibiting* facial reactions can dampen your reactions to emotional images.

Why might your expressions influence your emotions? One prominent explanation is that emotions themselves are *embodied*. This means that the act of thinking about an emotion is composed, partly, of subtle changes in a number of body states associated with that emotion. If emotions are embodied, they each, in part, correspond to a specific set of body states—many of which are neurophysiological and some of which are muscular. It's thought that some of these body states involve the motoric actions of the face. In this way, changing a face's motor actions actually changes a small part of the emotion itself. This could be why gently biting on a pen can heighten your level of amusement.

Regardless, it seems that not only are smiles contagious, but they also may serve to infect us with a subtle sense of well-being. And as with most contagions, the effects of smiles can occur without our awareness.

You Can Perceive Personality from Faces

NOT ONLY ARE EXPRESSIONS contagious, but they also can be chronic. Perhaps you've heard the adage that your life is written on your face. It has some scientific support. Research suggests that by the time you're in your seventies, strangers will be able to perceive something about your emotional life simply by looking at photographs of your face. By this age, your face will be better at conveying some emotions than others. The reason for this is simple: exercise. Underneath your skin, your face is composed of a nexus of over 50 different muscles interconnected in one of your body's most complex muscular arrangements. And as is always true of muscles, the more you exercise them, the better they work. Thus, the more you smile, the easier it is for you to use your face to express happiness. This becomes especially true as you get older.

This is why when strangers see photographs of your septuagenarian face expressing different emotions, they will find some easier to interpret than others. The ease with which your different expressions can be recognized will be related to your personality (as measured from personality questionnaires). If, for example, people have trouble knowing when your face is sad, there's a good chance you'd be the type of person who doesn't get sad, or express sadness, often. Expressing sadness infrequently may render the muscles used to express facial sadness slightly less effective than those used for other emotions. This would make your sad expressions somewhat less distinct and a bit less recognizable.

These lifelong *expressive habits* will also influence the overall structure of your aged face, not just your expressions. Again, the reason is exercise. The frequency and extent of your expressions can influence the size and shape of your facial muscles as well as how your skin wrinkles. These structural changes can influence the appearance of your face even when it is not expressing a particular emotion. This could explain why, as research has shown, strangers can interpret personality characteristics from older faces that are *neutrally* posed. For example, if strangers were asked to choose *some* emotion to describe a photograph of your neutrally posed 70-year-old face, they may say it looks "cheerful." There's a good chance, then, that you would have scored high on the "cheerful" dimension in a personality questionnaire.

So, personality characteristics and their expressive habits can mold faces. This may be one way you're able to read faces for clues about character. Despite your teachers' admonition not to judge people from their appearance, you do it all the time—you can't help it. And you can do so extremely rapidly. Research shows that you can establish an impression of a person from seeing his or her face for just one-tenth of a second: seeing the face any longer won't change this impression. And some of your impressions are likely to be correct. You *can* accurately judge aspects of someone's character from the face. This is true even if you were asked to judge personality traits from face photographs devoid of expression, makeup, and hair. You can even judge personality traits from faces that are "homogenized."

A group of faces can be homogenized by combining them into a single composite image. Typically, 5 to 15 face photographs are cropped, aligned, made semitransparent, and then functionally overlaid on top of each other using a digital averaging technique. This produces a single composite facial image that pretty much looks like a normal face. Yet the composite is homogenized in that it doesn't contain any one face's idiosyncratic features. Instead, it contains the average shape, features, and configuration of all the faces that make it up. This composite technique allows for tests of how *general* appearance and personality characteristics bear on our perception.

In one experiment, a composite face was created from the faces of 15 college students who tested high on an *extroversion* questionnaire. (All of the faces had a neutral expression.) When this composite face was shown to naïve subjects, they identified it as appearing more extroverted than another composite made up of faces from students scoring low on extroversion. Similar results have been found for the personality traits of conscientiousness, neuroticism, and openness to new experiences, at least for female faces. Other research using noncomposite faces has shown that raters are quite consistent, and relatively correct, in rating faces for their intelligence. And in a very recent study, naïve subjects were able to identify a composite face of individuals describing themselves as "open to short-term relationships and sex without love." One can imagine the evolutionary benefits of this skill.

It's unclear what makes a face look extroverted or neurotic or open to short-term relationships—especially when the faces have a neutral expression and are homogenized through the composite technique. It is known that a face's perceived attractiveness carries over to how we interpret its owner's personality characteristics, a fact you'll soon learn. Further, there may very well be a self-fulfilling prophecy component to how faces convey personality. If, for example, your face appears gregarious, the social feedback you receive from that fact may influence you to become more gregarious.

But part of how faces convey personality may come from how their expressive habits mold their look. A face may look extroverted because its expressive habits have subtly changed its structure, even when it's seen in a neutral pose. And your extraordinary perceptual skills with faces allow you to pick up on this structural information.

Still, your ability to perceive personality from faces should not be overstated. The composites in most of the demonstrations were made from individuals scoring at the extreme ends of a personality dimension. It's unclear whether you can detect more subtle differences between faces when seen homogenized and neutrally posed. So your teachers were probably right: Like for the proverbial book's cover, you shouldn't fully judge a person by his or her face. But as the research shows, it's not a bad place to start.

The molding of faces by expressive habits may help explain one last provocative facial fact. You look like your significant other. Moreover, you've grown to look *more* alike over time. It's long been known that people choose mates with whom they share some facial characteristics. This is a result of cultural, nutritional, environmental, and personality similarities between partners. Attractiveness is also correlated across partners, as you'll soon learn. But there is also evidence that spouses grow to look *more* like each other the longer they're together. There are likely a number of reasons for this, including common diet and climate. But it could very well be that converging expressive habits play a role as well. The idea is that because of the recurring empathetic expressions partners share, their faces, and particularly expressions, become slightly more similar over time. As support for this idea, couples who describe their relationship as happy, and report sharing worries and concerns, show a *greater* convergence in facial appearance than couples who don't describe their relationship this way. The more spouses share empathetic expressions, the more their faces converge in appearance.

So look in the mirror. If you've grown to look more like your partner over the years, it bodes well for your relationship.

You Can't Resist the Power of an Attractive Face

CHERYL TIEGS KNOWS SOMETHING about faces. Considered one of the first supermodels, her face was on the cover of major fashion magazines by the age of 17. After college she became the first model to appear on the cover of the *Sports Illustrated* swimsuit edition three times. In 1978 her face graced the cover of *Time* magazine, which dubbed her "The Great American Model." It's a time Tiegs remembers as being different for models. "Back then, we could maintain an element of mystery. People just knew us from how we looked. These days, the media are so out of control—every move these girls make is being followed. I feel bad for them. I used to love being mysterious."

But as her career progressed, Tiegs allowed herself to become less mysterious. She authored books, acted as a fashion editor for a morning news show, and developed her own fashion line for a major department store. These days, Tiegs continues her entrepreneurial endeavors with clothing and makeup lines. She also models and appears in movies and television. Much of her time is devoted to philanthropic pursuit and raising her son.

Her experience in the modeling and fashion world has given her a unique perspective on facial attractiveness. "When I first started out, thank God, they were looking for the blond, blue-eyed, thin-lipped California girl. Since then, I've seen preferences come and go. For a while, there was a preference for more exotic looks: thicker lips, darker complexions. And now, we're in a period where almost anything goes." But she does think that there is something common to all beautiful faces. "Classic features are classic features. And I can appreciate good features along with everyone else."

Tiegs is quick to add that attractiveness alone is not enough for a successful career in front of the camera. "I've seen *so* many girls with perfectly chiseled features come and go. Despite what people think, modeling really is hard. For any shot, you have to be completely aware of what every part of your body is doing. Your hip might need to be at an odd angle, your legs, arms, and even fingers need to be positioned just so, and you may be in high-heels, standing around lava.

"And then there's the part that many girls have problems with. You'll have to convey a *convincing* mood with your face. It's not enough to have 'a sexy look' or a 'sweet look.' There needs to be something behind the shot, and that takes concentration. For me, I'll think of something that helps my face take on an expression. If I'm asked to show love, or happiness, I'll think of my son Zack, and I'll feel my features soften. And this more natural expression definitely shows up in the photograph."

When I ask if her feelings on beauty have changed over the years, she replies, "I am really finding that beauty has nothing to do with age. I remember seeing Jessica Tandy [an actress active through her eighties] at the Academy Awards years ago. She was so poised, so comfortable— she was the most radiant woman at the Oscars. Coincidentally, the next day I ended up sitting near her on a plane and she was absolutely lovely. Beauty comes from the inside. The way a person behaves and carries themselves determines their attractiveness."

And this is no doubt true. A person's attractiveness is more than just a pretty face. But the research on facial attractiveness suggests that simple appearance does have a strong influence over us: stronger than most of us realize. And for those of us who look less pretty than, say, professorial, the results can be intimidating. Research has long shown that adults with greater facial attractiveness are more popular among their peers, have more dating and sexual experiences, gain greater occupational success (including 15 percent higher salaries), and are less likely to be found guilty in criminal trials. The more attractive also tend to have greater emotional and physical health. Children who are more attractive are treated more favorably by strangers, and even their parents! But children have their revenge: even as newborn infants they look longer at attractive than unattractive faces.

One of the most studied influences of attractiveness is how it affects mate selection. Over 30 separate studies have found strong correlations in attractiveness between romantic partners. These studies typically involve collecting attractiveness ratings on facial photographs (cropped, expressionless) of numerous individuals. The ratings turn out to be similar for romantic partners, especially those who are married. These effects seem to exist across all cultures in which the issue has been studied.

Certainly, physical attractiveness also depends on the body. But there is evidence that in judging overall attractiveness from photographs, both men and women base their judgments twice as much on faces than bodies, or any other visible characteristics.

Here are just a few of the more recent studies showing the overwhelming influence of facial attractiveness. First, to be an electable female politician, it helps if your face looks attractive as well as competent. In a recent study, subjects were shown photographs of female candidates running for the U.S. Congress. When asked which candidates they would likely vote for, subjects often chose those with more attractive, as well as more competent looking, faces (based on the ratings of another set of subjects). Another recent experiment revealed that individuals with attractive faces are more successful in door-to-door fund-raising than their less attractive counterparts. Finally, analyses of contestant behavior on a number of *Survivor*-type shows reveal that more attractive contestants most always fare better regardless of their relative skills and competence.

Of course, it is difficult for the real-world research to establish that it is facial attractiveness *alone* that garners such positive effects. There are myriad intervening factors that likely play a role. Individuals with attractive faces may be more likely to attend to grooming and hygiene, which no doubt influence dating and professional success. The self-confidence facilitated by facial attractiveness can have similar effects. In fact, a recent study showing that more attractive CEOs ran more successful real estate agencies also determined that these CEOs were more self-confident—a characteristic that also affects professional success. The complexities of real-world research make it difficult to evaluate the direct influences of facial attractiveness.

Alas, the more tightly controlled laboratory research reveals our same shallow tendencies. When shown images of cropped faces posing neutral expressions, subjects judge more attractive faces as belonging to individuals with more desirable personality traits. This tendency for us to think that what is beautiful is good is known as the *physical attractiveness stereotype*. It has been shown to apply to the faces of children, as well as adults, based on ratings of photographic, composite, and computer-generated faces. Based on ratings of these images, children with more attractive faces are assumed to do better in school, be better adjusted, and to have more social competence than children with less attractive faces. Adults with more attractive faces are assumed to be more professionally competent, more interpersonally skilled, and be better adjusted emotionally than those with less attractive faces.

Face . . . the sad facts.

You Know the Face of Reproductive Potential

Given the obvious sociocultural power carried by facial attractiveness, it's not surprising that your implicit perceptual abilities to evaluate it are formidable. You're able to detect attractive faces astonishingly quickly and without conscious awareness. Similar to your skills at recognizing facial identity and expression, you are able to "evaluate" the attractiveness of subliminally presented faces. For example, asked to rate the attractiveness of faces presented at one seventy-fifth of a second you'd have no awareness of even seeing faces and feel that you were guessing on every trial. Still, your attractiveness ratings would be similar to those you'd make when seeing the faces at longer, consciously accessible, durations. And if a subliminal attractive face was presented to you just before you saw a word with some positive connotation (*joy, pleasure, happy*), you'd categorize that word as "good" faster than if the subliminal face were unattractive. Your categorization of negative words (*nasty, evil, awful*) as "bad" would be faster if preceded with a subliminal unattractive face. (These effects don't work if the faces are presented upside down or if subliminal house images are presented instead of faces.) These results, you may recognize, are related to the *physical attractiveness stereotype*: it seems you can't help unconsciously evaluating the attractiveness of faces to the degree it affects the speed of your character judgments.

So *why* is facial attractiveness so important? One clue to the answer is another discouraging fact for those of us with a less pretty, professorial look: beauty is *not* in the eye of the beholder. To an intimidating degree, an attractive face is attractive everywhere. As Cheryl Tiegs has observed, despite fads for particular looks, there is something universal about pretty faces. In fact, there is a vast literature showing that many characteristics considered attractive in faces are culturally and historically universal. This fact, along with its sociocultural power, has led many scientists to conclude that facial attractiveness serves an important evolutionary purpose. Attractiveness may very well advertise an individual's genetic health and potential for healthy offspring.

Consider the aspects of attractiveness that seem universal. For example, across cultures and history, youthfulness is considered an attractive feature. Youthful characteristics such as large eyes, full lips, and smooth skin are known to have universal appeal. These characteristics are some of the most common goals of cosmetics. It is thought that a youthful appearance helps advertise fertility and the ability to produce viable offspring.

Attractive faces are also symmetric. Faces whose left and right sides approximate mirror images are rated more attractive than those with noticeable asymmetries. The preference for symmetry has been shown with photographic, composite, and computer-graphic facial images. Recall from Chapter 4 that body symmetry is considered an attractive aspect of most animals, whether conveyed through sight or smell. An animal's symmetry is thought to help advertise its genetic health including its resistance to disease. Symmetry also advertises the potential health of an animal's offspring. These facts may very well hold for human faces. We may find symmetric faces attractive because they belong to healthier individuals who can produce healthier offspring. In fact, there is evidence that women's preferences for symmetric male faces increase during their most fertile period.

Amusingly, our preferences for facial symmetry may play a part in the infamous "beer goggles" phenomenon. There's research showing that alcohol does lower our criteria for the faces we find attractive. When we imbibe, we're more likely to find a greater number of both male and female faces attractive. There are probably multiple besotted reasons why this occurs. But one less known effect of drinking is that we become less sensitive to asymmetries in shapes, potentially including facial shapes. It may make sense, then, to take a few facial measurements before making a date when under the influence.

Another dimension of facial attractiveness known to hold across cultures is gender distinctiveness (sexual dimorphism). The research clearly shows that men like female faces that appear more feminine. And, in general, women prefer men to have more masculine-looking faces, but this is tempered when they are asked about preferences for long-term versus short-term mates. Like facial symmetry, gender distinctiveness is thought to advertise genetic health. And like symmetry, women's preferences for masculine-looking faces increase during their most fertile periods. When women are looking for short-term relationships, they also prefer more masculine, symmetric faces—as well as, it seems, scars. When British women were shown facial photographs and asked who they found most appealing for a short-term fling, they preferred faces with minor scarring over the same faces without scars. Facial scaring might indicate bravery and higher levels of testosterone—a concept embraced by the Yanomamö of Venezuela. These tribesmen use makeup to accentuate the facial scars they sustain during ritualized fighting.

One other universal characteristic of attractive faces is that they are *average*. While this may sound paradoxical, it makes perfect sense once you understand what is meant by *averageness* in this context. Recall the composite face technique in which a group of cropped faces are aligned and then digitally superimposed. The result is a single facial image possessing the average characteristics of all the faces that make it up. It turns out that, in general, composite faces are judged as more attractive than the individual faces used to produce them. Also, the greater the number of faces used to make a composite, the more attractive the composite looks. It's thought that as the number of faces increases, the more the composite possesses features reflecting the *average* face of a population. And averageness is an adaptive trait: it reflects genetic normalcy. Thus, the averageness incurred in a composite makes it attractive for biologically sensible reasons: it essentially advertises the genetic health of its (virtual) owner.

In sum, the universal preferences for faces that are youthful, symmetric, gender distinctive, and average could be based on how these characteristics advertise reproductive potential. Certainly, this may be less relevant to modern society where the link between facial attractiveness and reproductive potential is diluted. Medical advances have offset many of the obstacles faced by less-than-optimal health and reproductive fitness—in vitro fertilization being just one example. And modern society provides myriad ways for attractiveness to be falsely advertised. Cosmetics, plastic surgery, and steroid supplements can give the face of most any individual a more youthful, symmetric, gender-specific, and average appearance.

Still, some provocative relations between facial attractiveness and reproductive potential persist. There is evidence, for example, that more attractive faces are associated with greater complexity in the major histocompatibility complex—MHC. As you learned in Chapter 4, greater complexity in MHC provides for stronger immune function in an individual and, potentially, his or her offspring. Greater male facial attractiveness is also associated with higher sperm quality. With regard to specific dimensions of attractiveness, there is a link between higher facial masculinity and better health in adolescent boys. There's also a relation between facial averageness measured at age 17 and childhood health for both men and women. Finally, it seems that in societies with less accessible medical care and greater incidence of infection, facial attractiveness in mates is more highly valued. While such sociocultural phenomena are complex, this could indicate that when needed, the attractiveness-reproductive health association plays a more important role.

It should be acknowledged that other explanations for the universal aspects of attractiveness have been considered. It's been argued that general principles of visual perception explain our preferences. With regard to symmetry, it's well known that we have an aesthetic preference for symmetry in *all* types of figures, not just faces. This could mean that our facial symmetry preference is based on the general workings of our visual systems, and not related to reproductive potential. Yet, the opposite could be true. Our general preferences for symmetry could be a by-product of our biologically driven attraction to facial and body symmetry. With regard to our preference for average faces, there is some evidence that this may be based, in part, on their apparent *familiarity* rather than an advertisement of genetic normalcy. An average face may be interpreted by your brain as familiar because it approximates the prototypic memory composite of all the faces you've ever seen.

However, the reported relations between actual health measures and dimensions of attractiveness are hard to ignore. They suggest that to at least some degree the information available in youthfulness, gender distinctiveness, symmetry, and averageness is there for you to implicitly use to select a healthy mate. It could be that general visual system principles help you to *attend* to the characteristics of symmetry and averageness.

Regardless, the sociocultural power of facial attractiveness, together with its universal attributes and automaticity with which it's perceived, suggests an important adaptive purpose to a pretty face. This is yet another fact troubling to those of us with a more professorial look.

But as even Cheryl Tiegs will tell you, attractiveness is more than a pretty face. "People ask me for beauty secrets all the time. I tell them what my first agent told me when I was finishing college. The secret to being attractive is to never stop learning—to stay curious and learn about everything. That enthusiasm for learning gets radiated—and that's beautiful."

You now know why Steve Preeg and Digital Domain's job was so hard. You are supremely sensitive to faces—and for good reason. A face conveys information about its owner's reproductive potential, genetic health, emotional state, linguistic intent, and, of course, identity. Face perception is such a core part of your perceptual repertoire that you incorporate subliminal facial information, and your own face inadvertently imitates the facial expressions you see. You use facial information whenever and however you can. Put simply, you are a face-perceiving expert and it was Steve Preeg and Digital Domain's willingness to take on this expertise that led to their Academy Award.

Impressionist Marilyn Michaels uses sense memory first to mimic the voices of celebrities and then to fully embody them.

CHAPTER 9

The Highest Form of Flattery

THE FUN ALL STARTED WHEN MARILYN MICHAELS WAS 11. "I would go around the house imitating the singers I heard on the radio. Patti Page, Sarah Vaughan, Connie Francis. By the time I was fourteen, I did my impressions for the owner of the Roxy Theatre in Manhattan. He was blown away and wanted to sign me to perform there. But my parents thought I should stay in school." Michaels's parents knew something about the commitments of being on stage. Her father was a singer with New York City's Metropolitan Opera and her mother was an actress in Yiddish theater, as well as one of the first female cantors in the country. Michaels's uncle was also a cantor and the preeminent Yiddish film star of his time.

But Michaels's talent for imitation ultimately won out. She matured into one of the best known and most versatile impressionists of her generation. She regularly appeared on the *Ed Sullivan Show* and *Tonight Show* in the 1960s and 1970s. Perhaps best known for her spot-on impressions of Judy Garland and Barbra Streisand, she performed on variety and game shows through the 1980s, and co-starred in a series devoted to impressionists, "The Kopycats." She is also well known for her performances on the Howard Stern radio show in the 1990s on which she helped Stern play tricks on his cohorts by pretending to call in as Zsa Zsa Gabor and Joan Rivers. Michaels is also an accomplished singer who has recorded many albums and has performed in theater. These days, she continues to perform as an impressionist and singer on radio and stage.

Michaels feels that her family background provided her with more than just vocal prowess. "My mother had a unique ability to imitate people. She was a natural mimic and did it for fun, socially. So I guess I get it from her. But also, my mother, father, and uncle were all actors. And acting is mimicry. In fact, it's the most important element. Think about those actors that are really able to lose themselves in a role. Meryl Streep, Marlon Brando, Laurence Olivier, Gary Oldman—they can immerse themselves in a part, and take on the characteristics of another person. They can take the sounds and dialects that they hear and know, and create an entirely new character."

Of course for a professional impressionist like Michaels, the trick is using acting and mimicry to re-create a character we all know. "For me, it has to do with sense memory: remembering in detail a voice I hear and then translating it vocally and physically. Most people can do some sort of impression of an uncle, or teacher you make fun of. But what makes the talent unique is to take those memories, and then create a *whole* individual, either someone famous like I do, or someone new, like Meryl Streep does."

So for Michaels, much of what distinguishes our casual mimicry from the master impressions she performs is her ability to embody a *full* character. This is consistent with findings from recent brain imaging research. In one experiment, fMRI imaging was conducted on the brain of British impressionist Duncan Wisbey as he performed 40 different impressions. Control subjects, with no formal impersonation experience, were also scanned as they performed the same impressions. Both Wisbey and the control subjects showed activity in the speech-related areas of their brains. But unlike those of the controls, Wisbey's brain also showed activation in areas associated with vision and motor control. This additional activation might very well be related to the fact that, according to Wisbey, he vividly imagines the people he imitates and then attempts to impose their features, voices, and mannerisms on his own body. Like Marilyn Michaels, his goal is to fully embody his characters.

It does take work for Michaels to successfully make an impression her own. "Depending on the celebrity, I'll watch multiple video or audio recordings of the person. Often I'll get the voice first, and then the mannerisms and expressions will follow. Sometimes though, there are characters that just don't register with me—they just don't capture my imagination or memory. But then I'll see someone else imitating the character, someone on *Saturday Night Live*, for example. Their exaggerations of a particular mannerism or sound will sit in my brain. And then I'll be able to start my own impersonation of the character and find other nuances. Sarah Palin [former Alaska governor and vice presidential candidate] is an example of a character I refined after first seeing her impersonated by Tina Fey on *Saturday Night Live*."

Michaels then practices the impression incessantly—sometimes in front of a mirror, or while videotaping herself. "When I was working on Sarah Palin, I practiced it over and over again. My husband was going crazy. But I needed to do this. I wanted to make her part of my consciousness. Now I can do her at the snap of a finger."

Much of what makes a successful impression is the right degree of exaggeration. It's the impressionist's talent to make a character immediately recognizable, and this comes from making a distinctive characteristic that much more conspicuous. Whether caricaturizing Liza Minelli's giddiness or Barbra Steisand's hand gestures, Michaels considers this a critical part of her craft. "We're all defined by these giveaway characteristics. And exaggerating the vocal and physical nuances of a celebrity is what makes an impression amusing. The more of these characteristics a celebrity has, the easier it is to re-create them."

Michaels has also been able to share her methods with other performers. "I taught Louise Fletcher [of *One Flew Over the Cuckoo's Nest* fame] to do Ingrid Bergman [*Casablanca*; *For Whom the Bell Tolls*] so she could get a part in the film *The Cheap Detective*. The part called for an Ingrid Bergman–type character. Louise is a fine actress, but she's from the Midwest, so sounding like Ingrid Bergman was a stretch for her. Here's what I taught her. Bergman was one of those actors who was carefully coached in English when she came over from Europe to make movies. Those actors were all taught to speak slowly and to be very conscious of their speech. If you watch those old movies, you can see the foreign actors working very hard to be understood. They were also taught to make their consonants very clear, especially those at the end of words.

If you listen to those actors, you can hear that the final consonants are very important.

"At the same time, Ingrid Bergman could never quite lose those Swedish *r*'s. That's often what sets a foreign accent apart—how the *r*'s are pronounced. It's what makes a Frenchman sound French, an American sound American, and a Russian, Russian. And it's what made Ingrid Bergman always sound Swedish. So before her audition for *The Cheap Detective*, I helped Louise learn to speak slowly, exaggerate her final consonants, and use Swedish *r*'s. She got the part."

■ ■ ■ ■ ■

WHILE SOUNDING LIKE INGRID BERGMAN might take the help of master impressionist Marilyn Michaels, sounding just a bit like the people with whom you interact takes no training at all. In fact, you do it all the time. On a largely unconscious level, you are always imitating people. You can't help it.

Whenever you interact with someone, you subtly imitate aspects of his or her mannerisms, speech, facial expressions, and behavior. Perhaps you've noticed yourself inadvertently imitating the accent of someone you're talking with. Every now and then, you may notice yourself standing with the same posture as the person you're talking to. But you subtly mimic all the time, and most of it occurs without your conscious awareness. And all this mimicking has a purpose. It facilitates your perceptual, coordinative, and social success. Research over the last 10 years shows that one of your most important implicit perceptual skills is your ability to integrate subtle information about other people—information that allows you to covertly and overtly imitate behavior.

You've already learned about your skills to implicitly mimic facial expressions. Covert facial mimicry allows you to better recognize the expressions and emotions of other people. But this type of imitation may very well apply to much of what you see—and hear—people do. And this imitation may have important perceptual and social functions.

It's been known for some time that we mimic the nonverbal behavior of people with whom we're interacting. Research in the 1970s showed that students tended to imitate the postures of the teachers with whom they were talking. Whether standing or sitting, the students were shown to lean in the same way as their teachers as they conversed. The students would also often cross their legs and arms if their teacher was doing the same. This imitation tended to increase as the positive rapport grew between student and teacher. A similar increase in posture matching has been shown for therapists and their patients over the course of treatment.

We also imitate each other's mannerisms. In a seminal study, Tanya Chartrand and John Bargh had each of their subjects sit at a table with another participant and describe a set of photographs. Unbeknownst to the subject, the other participant was actually an experimenter. As the phony participant discussed the photographs, he shook his foot in a seemingly inadvertent, unconscious way. Once all the photographs had been described, the phony participant left the room, and a new phony participant entered. Rather than shaking his foot, this second phony participant rubbed his face often, again in a seemingly inadvertent, natural way, as he discussed the photographs. To add to the overall sneakiness of the experiment, subjects were secretly videotaped during both phases.

Based on evaluations of the tapes, the results were clear: subjects spent much more time inadvertently foot shaking when they were with the foot-shaking phony participant and more time rubbing their face when with the face-rubbing participant. In both cases, subjects were unaware of their mimicking behaviors. Careful postexperiment interviews revealed that subjects did not know that either they or their partners were foot shaking or face rubbing. Further, the mimicking occurred even though little rapport had been established. As part of the experiment design, neither of the phony participants made eye contact with the subjects, and one of the two never even smiled.

Chartrand and Bargh's study is considered one of the first experimental validations of the *chameleon effect:* our unconscious tendency to mimic the postures, mannerisms, and facial expressions of other people. Since then, other types of chameleon effects have been observed. For example, not only do you adopt the types of mannerisms you see others perform, you also adopt their speed. Imagine being in this experiment. You're told that you will see a series of actions performed by a "point-light actor." You've already learned about point-light images of a face, in which motions are conveyed through the movements of illuminated points against a black background. The point-light actor technique also shows points moving against a black background, but the points are typically larger and are applied to the 12 major joints on the body (ankles, knees, hips, shoulders, elbows, and wrists). Despite the reduced nature of these images, you'd be able to determine what action is being performed (walking, hopping), the gender of the actor, and even the identity of the actor, if it were someone you knew.

In this particular experiment, your job is to watch a series of two-second video clips of a point-light actor performing 26 actions (walking, running, kicking) and to recognize each. But in between seeing each action and naming it, you are to perform a very simple distracter task. This task involves looking at a small cross (+) presented in the middle of the video screen and pressing one of two buttons depending on whether its vertical or horizontal component is dimmer. Once you perform this task, you can then name the point-light action you saw a few seconds earlier.

There's one last twist. To make the point-light actions more challenging to recognize, the video clips are sometimes shown at twice their normal speed, sometimes at half their normal speed, as well as at normal speed.

How do you think you'd do at recognizing the different point-light actions? Pretty well it turns out, regardless of the presented speed of the video clips. But that's not important. What's important is how you'd perform on that boring distracter task: the one where you pressed the buttons to indicate which piece of the cross was dimmer. It's an easy task, and you'd almost always be correct. But the *speed* with which you'd respond would depend on whether the preceding point-light action was seen at double, half, or normal speed. The faster the point-light action, the faster you would respond on the cross task. And you would have no awareness of this influence during the experiment. Your response rate in the cross task would inadvertently mimic the rate of the point-light action.

Critically, you must be seeing *human* actions to induce your response rate mimicking. If the task instead involved identifying the motion of boxes seen at normal, double, and half speeds, no influence would occur. The fact that inadvertent imitation is most effectively induced by human behavior has been observed again and again. For example, you are faster at pressing one of two buttons if the target button is indicated by a video of a finger pressing that button than of a dot appearing over the button. The importance of a human component to induce inadvertent imitation will be explained soon.

Your imitation of rate extends to an action you rarely think about:

your breathing. Research suggests that even while sitting comfortably in a chair, your respiratory rate would increase as much as 30 percent if you watched a video of someone vigorously exercising. This would be especially likely if you were watching someone run on a treadmill with an increasing speed. Perhaps just going to the gym to *watch* someone work out is better than not going at all.

But if exercising by proxy is still too much trouble, you have another imitative option for keeping fit. You can dine with light eaters. Research suggests that you inadvertently imitate the eating behavior of the people you're with, eating more or less depending on how much they eat. This research is generally conducted in laboratory settings and involves eating and interacting with strangers: not the most natural context to reveal your true eating behavior. But imitative eating has been observed in multiple experiments and regardless of whether subjects are sated or food-deprived for 24 hours. And in all cases, subjects were unaware that the amount they ate mimicked that of another person.

■　■　■　■　■

YOU'VE BEEN IMITATING ALL your life. As you learned, by the time you were two weeks old, you imitated some facial behaviors. These behaviors included blinking, sticking out your tongue, and facial expressions approximating emotion. At that time, you also imitated facial postures related to speech articulations. In fact, you did this whether you were looking at a face or just *listening* to someone talk. Whether your eyes were opened or closed, you would often form an *m* with your lips when listening to someone make an *m* sound, and an *ah* when you heard an *ah* sound. So as early as they could be tested, your imitative abilities were multisensory.

Of course your imitation skills of heard speech have been refined since then. Besides inadvertently imitating a person's accent, you partially imitate the speech rate and intonation (the pitch and loudness changes). To some degree, you also mimic word choice and word order (syntax). More striking is that when it comes to fine-grain aspects of speech, your inadvertent mimicking surpasses your conscious control of your articulators. Consider the very subtle difference between how you articulate *bat* and *pat*. To understand this difference, put your hand just in front of your mouth and say each. You'll feel that there is a breath of air expelled in the beginning of *pat*, but not *bat*. The air is expelled because of a slight delay between when your lips part for the *p* and when your vocal chords start vibrating for the *a*. The air rushes past your nonvibrating vocal chords and then out your mouth. This is true whenever you make a *p* sound.

For *b*'s on the other hand, your vocal chords start vibrating just before you part your lips. Your vibrating vocal chords capture the rushing air and this is why you don't feel the air burst when you produce *bat*. So while for *p* there is a small delay—about one-twentieth of a second— between when your lips part and when your vocal chords start vibrating, for *b* there is no delay. And this is what largely distinguishes these two consonants.

So what does this have to do with imitation? It turns out that when you produce a *p*, the *extent* of this very tiny delay between lip parting and vocal chord vibrations depends on whom you're talking with. Different talkers have slightly different voice delays in their *p*—ranging from about one-twentieth to two-twentieths of a second. The length of the delay depends on a talker's native language and regional dialect, as well as physiology. And when you talk with someone, you often partially mimic the way he or she delays the vocal chords when producing *p*. (In fact, the same is true of *t*, *k*, and other "voiceless" consonants that involve a delay between moving upper articulators and starting vocal chord vibrations.)

What's most interesting about this fact is that in inadvertently mimicking someone's vocal chord delay, your own delay can systematically shift as little as *one two-hundredth* of a second. There is no way you could *intentionally* produce a *p* delay that shifted just that much. It's just too subtle. This means that the inadvertent speech imitation that comes with talking to someone imparts more fine-grain control over your articulators than you do.

INADVERTENT IMITATION MAKES YOU A BETTER PERSON

So, WHY DO YOU INADVERTENTLY imitate? There are probably many reasons, ranging from a fine-tuning of simple perception and action to a basis for your language acquisition. But one of the best understood reasons is imitation's social importance. Imitating others and *being imitated* facilitate your interactions. This was demonstrated in the lab in another experiment by Chartrand and Bargh, using the same phony participant tricks. In this experiment, the phony participant subtly mimicked the behavior of a *subject*. This imitation was carefully choreographed. The phony participant adopted the subject's postures slowly and smoothly, and a few seconds after the subject performed them (all secrets to inconspicuous imitation). When being covertly mimicked in this way, the subjects found the participant more likable and judged the interactions as having gone more smoothly. This occurred despite subjects never being aware of the imitative behavior.

Numerous other studies have revealed similar social benefits. Being subtly imitated by a "negotiator" would make you more likely to agree with his or her opinion. Being imitated by a pretend cola salesman would make you more likely to rate the soda favorably and drink more of it during your interaction. You'd give a higher tip to a waitress who imitates your order verbatim, rather than paraphrasing. And you'd even rate a computer-animated "interviewer" as more persuasive and positive if it subtly imitated your own head nodding. Finally, when GPS driving directions are conveyed by a voice that matches your own vocal emotion, you're less distracted than if provided directions by a voice not matching your emotion. In all of these examples, you'd have little if any conscious awareness that you'd been mimicked in any way.

But imitation doesn't just facilitate your positive behavior toward the mimic, your generosity generalizes. Imagine being in this experiment. You walk into an office and sit in front of a desk, behind which sits an experimenter. The experimenter tells you that your task is simply to describe and rate a series of magazine advertisements she shows you. As you describe the ads, she subtly imitates your postures: whether you're leaning forward or back, whether your arms are folded or opened. You have no idea you're being mirrored. After about five minutes of this, she tells you that for the second part of the experiment, another experimenter will lead you through a different task. She then leaves the room, and the new experimenter enters. But as this second experimenter walks toward the desk, she "'accidentally" drops a bunch of pens.

Would you help her pick them up? You'd almost have to. Because you had just been imitated by the first experimenter, you'd be *twice as likely* to help with the pens than if you hadn't been imitated. Being subtly mirrored by one person facilitates positive behavior toward another. In a related experiment, subjects who had been imitated were 80 percent more likely to donate their subject participation money to charity. Imitation isn't just the highest form of flattery, it makes you a better person.

The social importance of imitation is also evident by observing when it is most likely to occur. Research suggests that you perform more inadvertent imitation if you're motivated to have a cordial interaction. This would be true even if that motivation were provided by subliminal words (*partner*, *together*) presented to you before the interaction. You'd also perform more inadvertent imitation if a very recent interaction hadn't gone well. If, for example, you felt left out of an online game, you'd unknowingly perform more imitations in your subsequent live interactions. There's also evidence that you do more imitating of individuals whom you believe have more in common with you socioculturally.

Besides showing the myriad conditions under which we imitate, these examples demonstrate that even while much of our imitation is inadvertent, its frequency and magnitude are influenced by social factors.

IMITATION MAKES YOU A BETTER PERCEIVER

SOME BENEFITS OF YOUR IMITATION are more basic than its function as social glue. It likely helps you develop and refine your perceptual abilities. Whether improving an athletic skill or learning a new language, the imitation you perform unconsciously is as important as that you perform willfully. And this imitation benefits your perceptual skills as much as your motor skills. This is, in a way, intuitive. If you are a regular tennis player or golfer, you may be aware of yourself watching these sports more closely, and with more perceptual sensitivity, than individuals who do not regularly play. But research is showing that learning even the most unusual motor skills, through relatively little imitative practice, can enhance your perceptual skills.

Imagine this. You sit in a chair, you're blindfolded, and then you're asked to learn a little dance. It's a boring little dance that involves only your arms, but it takes a few sessions for you to master. Essentially you're taught to slowly swing your arms back and forth with a strange inter-limb timing. Rather than having your arms swing in completely opposite directions as they do when you walk, you're taught to wait until one arm is swinging down a few degrees before changing the direction of your other (your arms swing at 270 degrees out-of-phase). You are taught this movement by being guided by the experimenter.

You master the arm dance after a few sessions and then begin the second half of the experiment. Your blindfold is removed and you're asked to watch a series of animated point-light actors walking in place, as if on a treadmill. As with all point-light stimuli, you see only a set of points move, nothing else. The walkers are shown to you successively in pairs, for about one second each. Your job is simply to say whether the walkers in each pair are walking in the same or a different way. The problem is, most of the walkers look drunk. At least they're not walking in the normal way. In fact, they've been animated to walk so that their opposite side limbs are oddly coordinated. This makes the task of judging the similarity of the walkers extraordinarily difficult. And overall you'd be pretty bad at it.

But there would be one type of odd coordination that you could perceptually discriminate pretty well. You'd be disproportionately better at perceiving the strange coordinative style you had learned when you were arm dancing. You'd be better, in fact, at making that discrimination than individuals who did not learn the dance, or who learned a *different* arm dance (230 degrees out-of-phase). (These latter individuals would be better than you at perceptually discriminating their arm dance.) Recall that you were blindfolded while you were learning your arm dance. This means your enhanced perceptual skills were not based on visual experience. Instead, your imitative *motor* experience enhanced your perceptual ability to recognize a particular motor act.

Evidence for motor involvement in your perception skills is also evident from how *constraining* your actions inhibits your perception. Recall that constraining the covert imitative actions of your face—by biting a pencil or by transient brain lesions—made it harder to recognize some facial expressions. Analogous effects have been found with other actions. If, for example, you were asked to determine whether the arm positions of an actor shown in two consecutive photographs were the same or different, you would perform less well if your own arms were occupied with an unrelated movement. Your judgments of the actor's *leg* positions would not be affected, however: that would only happen if your legs, but not arms, were occupied with different movements.

Next, your ability to visibly judge the walking speed of a point-light actor would be greatly hindered if you, yourself, were walking at a different speed. And your visual judgments of the weight of a small box you see lifted by someone else is affected by simultaneously lifting a box yourself (lifting a light box makes the weight of the box you see being lifted look heavier, and vice versa). This is also true for your body's weight. Your predictions of how high another individual can jump is lowered if you first walk around with weights on your own ankles for five minutes.

The same sort of influences can occur when you're listening to speech. Imagine being in this menacing experiment. You sit in a chair, and wires are connected to the two corners of your mouth with medical tape. The wires each extend away from your face, where they're attached to robot arms. You're told that you'll listen to simple words over headphones and that the robot arms will pull randomly on the corners of your mouth—lightly, you hope. Further, the robot arms can move all around your head, allowing them to pull your mouth in multiple directions. Sound fun? No, *sounds different*. Depending on how your mouth is stretched, the words you hear will change slightly. For example, if you were played a series of words synthesized to sound somewhere between "head" and "had," you would be more likely to hear these words as "head" if your mouth corners were simultaneously pulled up with each presentation of the words. If, on the other hand, your mouth corners were pulled down synchronously with the words, you'd more likely hear them as "had." This makes sense because articulating "head" versus "had" involves mouth corner movements that subtly shift up and down, respectively.

If these sinister methods don't appeal to you, no worries. You can induce the same sort of effect by silently articulating one syllable while listening to another. If, for example, you silently articulate *ga* while listening to *pa*, you'd report hearing *ka* much of the time and *ta* other times. You'd rarely report hearing the actual *pa* you were presented over headphones.

In all of these examples, occupying your motor system influences your perception of another person's actions. And, these effects are motor system–specific: occupying your arms specifically affects your perception of arm movements; occupying your mouth changes your perception of speech. This suggests that like for facial expressions, you use your own motor system (including its muscle-tendon sense feedback) to help you perceive what someone else is doing.

■ ■ ■ ■ ■

BUT WHAT OF *actual* covert movements during perception? Recall that when you see a facial expression, you perform barely perceptible imitative movements of that expression yourself. While very, very subtle, these movements can be measured by recording the activity of your facial muscles or, in some cases, by closely watching your face change. Does the same thing happen when you watch someone perform other types of actions? When, for example, you watch someone's fingers grab something, are your finger grabbing muscles subtly active? Well, not to the degree that they can be observed by eye or with straight muscle recordings. But there is a way to record sympathetic muscular reactions, and it involves the technology used to induce transient brain lesions.

To create transient brain lesions, transcranial magnetic stimulation— or TMS—is administered to a brain area repetitively and intensely. But it doesn't have to be used like this. If administered as a single, weak pulse to the parts of the brain that activate muscles (motor cortex), TMS can functionally prime your muscles for movement. Applying light and short stimulation in this way creates measurable electrical activity in the muscles of your extremities (more specifically, the extremities on the opposite side of your body from the side of the brain stimulated). This muscle activity (motor-evoked potentials) is typically recorded with a small electrode placed on the skin over a muscle.

Now comes the interesting part. It turns out that the amount of muscle activity recorded when the pulse is administered depends on what you're *looking* at. If, for example, you're looking at an experimenter grasping an object when your brain is pulsed, greater activity would be recorded in the muscles you use for grasping. If, instead, you were watching the experimenter perform an arm movement, more activity would be recorded in your arm muscles. These effects would occur despite the fact that, as far as you knew, you were sitting perfectly still. And these empathetic muscle reactions seem to follow the *steps* of the seen action. By recording from the muscles on the inner thumb and the side of your index finger next to your thumb, the relative activity of these muscles would be shown to change in a grasp-appropriate way—and in synchrony with the phases of a seen grasping action.

Your pulse-induced sympathetic muscle activity can also be changed by *listening* to someone's actions. Listening to manual actions such as paper ripping or typing would increase the activity of your index finger muscle. Listening to footsteps or thunder sounds would *not* have this effect. And the same type of effects occur when listening to speech. Hearing syllables containing tongue-tip movements (*ta, da*) increases activity in pulse-activated tongue, but not lip, muscles. In contrast, hearing syllables containing *lip* movements (*pa, ba*) increases activity in lip, but not *tongue*, muscles.

So, with a little help from TMS pulsing to your motor brain, your muscles *can* be seen to covertly imitate the actions you see and hear. And occupying your muscles with unrelated actions can hinder your perception of other actions involving those muscles. This research suggests that your covert imitation can help you perceive actions. It should be noted, however, that as for perception of facial expressions, your covert imitation isn't *necessary* for perception of other's actions. You obviously *can* recognize arm and hand actions when your arms and hands are occupied with other tasks. And the influence of performing unrelated actions on perception of someone else's actions, despite being muscle-specific, is fairly subtle.

So, the influence of your covert imitation on your perception may be best understood as facilitative, rather than necessary. But your covert imitation may have other purposes as well. And to understand those purposes, it helps to know some background on the discovery of something very surprising about your motor brain.

Your Brain Is Imitating All the Time

HISTORICALLY, THE PARTS OF YOUR brain responsible for action planning (premotor and parietal cortexes) have been considered mostly as one-way issuers of commands to coordinate and control your body. While certainly influenced by other brain regions, these motor planning areas had been thought to work on the output side of how your brain channels information. All that changed in the mid-1990s.

It was one of those wonderful scientific serendipities. Giacomo Rizzolatti and his colleagues at the University of Parma (Italy) were testing how cells in the motor area of a macaque's brain were involved in planning hand and mouth movements. They were recording from cells as the monkey was grasping raisins and placing them in its mouth. And then, as the monkey was taking a break, one of the experimenters reached for the raisins. Rizzolatti and his colleagues were completely surprised by what occurred. Despite the fact that the monkey was not reaching, its motor cells were reacting as if it were. When the monkey watched the experimenter reach for the raisins, parts of its motor brain were responding as if it were performing these reaching actions itself. Monkey see, monkey brain do.

Rizzolatti and his colleagues later observed that these cells responded only when seeing actions involving the body parts that they helped control. Cells active during the initiation of hand movements were responsive to watching another individual (monkey or human) perform *hand* movements. Cells involved in initiating mouth actions were responsive to seeing mouth actions. For these reasons, Rizzolatti and his colleagues labeled this class of cells "mirror neurons."

We now know that there are different subtypes of mirror neurons, some responsive to very specific details of actions (thumb and index finger grasping an object), while others are more broadly receptive (any part of the hand grasping the object). Many mirror neurons seem to respond to multisensory information for perceived events. The cells involved in initiating hand tapping actions are responsive to both the sight and sound of hand tapping. Also, most mirror neurons will only respond to a hand grasping toward an *object*, and not a hand simply pantomiming this action. At the same time, however, if a seen object is then obstructed by a curtain, observing a grasping action that moves behind the curtain still induces a mirror neuron response. This has been interpreted by some researchers as evidence that the cells can "infer" the presence of a target object. These presumed inferential characteristics of mirror neurons have led some researchers to believe that the neurons don't simply register a perceived action, they are instrumental in understanding another animal's *intention*. This is a controversial idea, however.

But what about you? Do you have mirror neurons? It is, of course, much more difficult to conduct recordings from individual brain cells in humans. But brain imaging (fMRI, PET) has been conducted to look for regions that are active during both initiating and observing actions. This imaging shows that regions you use for motor planning (parietal, premotor) *are* responsive when you watch actions such as hand grasping. And more precise imaging has revealed that distinct sites of your motor brain (in the premotor cortex) active during *specific* performed actions—grasping, chewing, kicking—are also the sites responsive to seeing those specific actions. Along these lines, imaging has also revealed that your areas involved in producing facial expressions are responsive to seeing the same facial expressions.

There is also evidence that *disrupting* specific motor areas of your brain can inhibit your perceptual skill. Recall that inducing a transient brain lesion (with transcranial magnetic stimulation) in the motor-relevant areas of your brain inhibits your ability to visually recognize facial expressions. Analogous findings have been observed for other actions. If, for example, a transient lesion were produced in the brain area you use to plan hand actions (left inferior cortex), your visual judgments of the weight lifted by someone else's hand would be disrupted. Relatedly, a transient lesion induced in another of your motor planning areas (premotor cortex) would disrupt your ability to recognize (quickly match) hand, arm, and leg *actions*, but not to recognize the body parts themselves. Finally, remember that task where you were instructed to push one of two buttons by either watching a video of a finger pushing the target button or seeing a dot appear over the button? If a transient lesion were induced in another motor planning area (Broca's region), your responses to the finger pushing, but not dot cuing, would be disrupted. All of these effects may be a result of transiently lesioning mirror system brain areas considered to be involved in both the execution and recognition of actions.

Thus, many researchers do believe that you have a mirror neuron–like system. And if you do, it's likely involved in many of your inadvertent *imitative* behaviors as well. In fact, brain imaging shows that while these areas are active during initiating and perceiving specific actions, they are *most* active during your actual imitation of actions. This could also mean that your mirror system has been an important part of your perceptual-motor development and has been involved in your acquiring all types of skills. It may also help underlie your inadvertent mimicking of faces, mannerisms, body language, and speech: all functions facilitative of your social interactions. With regard to *covert* imitation, your mirror system could help prime your body for imitation, a fact supported by the TMS-pulse priming of sympathetic motor actions discussed earlier.

And, as intimated in the previous chapter, your mirror system areas seem to respond to *human* actions, and not to the same actions performed by a robot. This may help explain the uncanny valley effects. It may take your mirror system's involvement for you to notice and strongly react to the uncanny aspects of quasi-human faces and bodies.

So your mirror system may help you be a better perceiver of others' actions. It may do this by helping you to see others' behaviors based on how you might perform them yourself. In being active in both an action's initiation and recognition, the mirror system could help provide a direct, automatic way for your brain to translate perception into action; and what you see people do into how they did it.

■ ■ ■ ■ ■

BUT SPECULATIONS ON THE importance of the mirror system have gone well beyond this. This often happens in psychology and neuroscience, after an exciting, unexpected discovery. Speculation on the breadth of a new discovery's implications can become, shall we say, ambitious. But while these speculations on mirror systems must be taken with a grain of salt, some interesting supportive evidence is worth noting. For example, it's been thought that dysfunctions in the mirror system may be responsible for some social skill deficits, including autism. Recall the speculation that the mirror system may help provide us with an understanding of another's *intentions*. It turns out that individuals with an autism spectrum disorder show difficulty with social skills often including interpreting others' intents. Depending on the severity, these individuals also often show deficits in imitative behaviors (including contagious yawning). In fact, recent brain scans have shown that many autistic individuals *do* have structural abnormalities in their mirror systems (premotor and parietal areas). Other brain scan research has revealed that autistic children show less mirror system activation when looking at, or attempting to imitate, facial expressions. Still, autism spectrum disorders are considered very complex and mirror system involvement in its causes is still speculative.

It's also been thought that your mirror system might be involved in your ability to feel empathy for people. Empathy is an emotion that requires understanding someone else's intentions. There is, of course, much more involved in empathy than the level of intention understanding your mirror system ostensibly provides. Still, some provocative new brain imaging evidence has shown that the degree to which we rate ourselves as "empathetic" correlates with the amount our mirror system reacts to watching simple hand gestures.

Finally, there is speculation that the mirror system may have served an important purpose in the evolution of language. For language to be successful, the sender and receiver must tacitly agree on the components of the language code. Assuming that the components of human speech are essentially the articulatory gestures forming speech sounds, then a mirror system would provide a simple way for the brain to construe the components the same for sending and receiving the message. You'll learn more about this idea in the next two chapters.

Regardless of whether the mirror system is involved in these functions, a few facts have become clear. The discovery of motor cells that respond to seeing (or hearing) the same actions they help initiate is exciting, and potentially important. In fact, the mirror neuron system has become one of the most studied issues in modern perceptual neuroscience. It is too early to determine the importance of the mirror system to our everyday behaviors or whether it plays a necessary, or even facilitative, role. Perhaps what is more important is the fact that the mirror system concept has helped motivate an awareness of our vast imitative tendencies. We now know that human imitation, whether intentional or inadvertent, whether overt or covert, is far more prevalent and important than we ever imagined.

YOU ARE STRONGLY DRAWN TO IMITATE SOMEONE'S GAZE

"How's your salmon?" Dave Thorsen asks me.

"Really good, thanks. And how's your—." But before I can ask about his pasta, Thorsen interjects.

"Then watch this." Thorsen, who is sitting just across the table from me, pulls out a half-dollar coin. He uses the thumb and index finger of his left hand to hold up the coin just between our faces. "If you take a simple coin like this and place it in your hand . . . ," he places it in the palm of his right hand and closes it. He then passes the coin to his left hand and says, "Then it stays there." Thorsen opens his left hand to show me the coin is there. He then takes the coin in his left-hand thumb and index finger again. "But if you do the same thing again," he places the coin in his right-hand palm, "and just hold it there . . . ," he waits a few seconds before moving the coin to his left hand, "then," he moves it to his left hand, "it disappears." He opens his left hand and then his right hand to reveal that the coin is in neither. The coin disappeared.

"Wow!" I respond. "That worked on me—and I'm a perceptual psychologist, and we're talking about the science of magic! Never mind that you made it vanish one foot away from me. That's amazing."

"*That's* misdirection," Thorsen replies.

I'm having dinner with Dave Thorsen at the Magic Castle in Hollywood. The Magic Castle is one of the world's premier private clubs for magicians and is the home of the Academy of Magic Arts. Thorsen is a charter member of the Castle and has been its primary magic instructor for over 15 years. He's in his early eighties, but seems much younger because of his full head of white hair and energetic demeanor.

"I teach my students three types of misdirection," Thorsen continues. "The point of each of them is to get the audience volunteer to look at or think about something *other* than what you don't want them to notice. You can use what we call 'patter.' You talk or ask them a question so they're distracted. Hopefully this will also get them to look you in the eyes. Another misdirection technique is motion. People's eyes follow motion, so if you slowly, prominently move one hand, you can use your other to do something that won't be noticed. Finally, there's gaze. People have a tendency to look where you're looking. So, for example, looking at your distraction hand is another way to get them to look away from what you don't want them to see."

I tell Thorsen that the misdirection he describes is exactly what's been observed in the research on the perception of magic. I then ask whether this emerging "science of magic" is a concern for magicians—whether he's worried that his tricks will be revealed.

"I have mixed feelings about all that," Thorsen replies. "For instance,

I *love* teaching magic. But I tell my students there's one part I don't love: giving away the tricks. And I teach them that they shouldn't give away the tricks, if they can help it. But at the same time, I think that magicians realize that even after a layperson learns the secret of a trick, they don't necessarily remember it the next time they see it. And magicians like to be creative with their tricks. People often don't know that they're seeing the same type of trick more than once because of each magician's unique interpretation. This is probably true when people learn about the science of magic.

"Still, I teach my students to never perform a trick twice to the same audience, and to keep the tricks coming. You don't want your audience to have time between tricks to think too hard about how a trick was done. A trick shouldn't be a puzzle, it should be magic."

Later that evening, Thorsen invites me to sit in on one of his classes. He's teaching to about 20 students who seem to come from all walks of life: physicians, gardeners, lawyers, hair stylists. But all are formally dressed, an entry requirement of the club. He teaches them a simple card trick involving patter misdirection. Before he starts he says, "Once you know this trick, you're going to be worried that it's so simple, no one will ever fall for it. But this trick always works, trust me."

He performs the card trick on a volunteer with the critical eyes of 20 students watching. And he's right, the trick works on all of us. He then teaches us how the trick is done, and it does seem especially simple. "The trick," he says, "is in how you move the cards as you're talking to your volunteer." He shows us a very simple motion involving placing one half of the deck on the other. "Now break into pairs and talk to each other as you try that motion over and over." We do this motion repeatedly for about 10 minutes. It turns out to be a very simple *looking* motion that takes substantial practice to do well. It would seem that for misdirection to be effective, a careful coordination is required between seemingly effortless overt and covert movements. The magic emerges from the choreography.

But of course the magic also emerges from the audience. And science is just beginning to understand how. Consider, for example, the misdirection technique of leading eye gaze. Recent research is showing how you are strongly drawn to imitate the gaze of another person. Some of this research actually uses the techniques of sleight-of-hand artists.

Imagine being in this experiment. You walk into a laboratory room and sit in front of a table. On the other side of the table sits a "magician," you are told, who will perform a trick. Before the magician begins, another experimenter places an odd little cap on your head. The cap contains some wires and a little mirror on its visor. It's been designed to allow the researchers to track your gaze. You're then told to watch the magician perform the trick and try to figure out how it's done.

Here's what you see. The magician picks up a pack of cigarettes and a lighter. He puts a cigarette in his mouth and brings the lighter toward its end. But just before lighting the cigarette, he looks down and realizes that he has put the wrong end in his mouth. He takes the cigarette out, turns it around, and puts it back in his mouth correctly. He then brings his right hand, which is holding the lighter, back toward the cigarette. However, to his apparent surprise, the lighter has mysteriously vanished! He shows you this by slowly opening his now-empty right hand. He then looks down at his left hand and opens it. You realize that the cigarette has also vanished: it is no longer in his mouth or in either of his hands. Voilà!

How did he do it? For the answer, let's follow your eyes. When the magician begins the trick, you first look at his face. When the magician looks down at his hands to pick up the cigarettes and lighter, your eyes also look down at his hands, about a half second later. Your gaze continues to follow his hands as he brings the cigarette to his mouth. When the magician then looks at the cigarette's end to notice that it needs to be turned, you also look at the cigarette, as well as his left hand as he uses it to reverse the cigarette. And here comes the magician's first secret action. While you're looking at his left hand reverse the cigarette, he inconspicuously uses his *right* hand to drop the lighter into his lap, under the table.

After the drop, the magician then looks down to his right hand which ostensibly holds the lighter. About a half second later, your gaze follows his so that it also looks at his right hand as he brings it up to light the cigarette. The magician then notices that the lighter has disappeared from his right hand, and shows you this by bringing that hand out to the side and slowly opening it. All through this motion, he continues to gaze at that hand. Your gaze again follows his so you can see that the lighter has, in fact, vanished. And as this is happening the magician performs the second secret action. About a half a second after he starts to bring his right hand to the side, he uses his left hand to inconspicuously take the cigarette from his mouth and drop it onto his lap under the table. You're watching his right hand open as this occurs. The magician then looks down at his left hand, waits a half second for your gaze to catch up, and then opens this hand to reveal that the cigarette has also vanished.

The trick to the trick is in how the magician uses his gaze (and pronounced hand motions) to draw your gaze to follow along. As Dave Thorsen's students well know, this is one of the most effective methods of misdirection. And research shows that gaze following can work whether or not you've been told to expect a trick. But timing is crucial. The magician intuitively knows that your gaze will follow his with about a half-second delay. So he does not perform the critical dropping actions until your gaze has had a chance to catch up.

What magicians have intuited forever, science is now validating with research: it's hard for you to resist the pull of another's gaze. While a person's gaze might not always induce your eyes to physically follow along, it will likely influence your reactions and attention.

Imagine being in this experiment. You sit in front of a computer monitor, and again slip on an eye-tracking cap. You see a very small black square positioned in the middle of the screen, flanked by a larger black square on each side. Your job is simply to watch the small, central square carefully, and if its color turns to blue, quickly shift your gaze to the large square on the right. If, instead, the smaller square turns orange, you're to shift your gaze to the large square on the left. So you sit back, fixate on the small, central square, and get ready to shift your gaze.

The central square turns blue. But just as you're starting to shift your gaze to the right, up pops a photograph just where the small, central square had been. The photograph shows the front of a woman's face with her eyes prominently gazing *toward the left*. Still, you complete your gaze shift to the right. The woman's face then disappears, and the small black square reappears, telling you its time to start the next trial. That was a practice trial. You're told by an experimenter that you will see this face appear on each trial, but you should concentrate only on the color-changing small square and disregard the face completely. You are then presented 360 trials like this. As you go through the trials you realize that the woman in the photograph sometimes gazes in the opposite direction from which you've been instructed, sometimes gazes in the *same* direction toward which you've been instructed, and sometimes gazes straight ahead.

It's a very simple task, and for the most part, you'd be pretty good at it. On the majority of trials, you'd shift your gaze toward the flanking square to which you've been instructed. But you would make some mistakes and shift your eyes in the wrong direction. Those mistakes would be significantly more likely if the woman in the photograph gazed in the opposite side from where you were instructed. And even when you were correct, your gaze shifts would be slower for this opposite gaze condition. It seems that even when told to ignore someone's gaze, it can still influence your own gazing behavior. Interestingly, these effects would *not* occur if arrows appeared in place of the gazing face, pointing in either the same or opposite direction from where you were instructed to gaze. This suggests that the mere suggestion of direction is not enough to influence your gaze behavior. Instead, it's the effect of seeing another *person's* gaze that induces the subtle but systematic influence on your own.

And it turns out that you don't even need to *consciously* see another's gaze for it to influence yours. If a similar experiment were conducted, but included presentations of a *subliminally* presented gazing face (shown for just one five-hundredth of a second) the same influences would occur. It seems that your gaze can be influenced by gazing faces you don't even notice.

And not only can another's gaze influence your own, it can also influence your opinions. Imagine being in this experiment. You look at a computer screen on which appears a small cross in the center. After about half a second the cross is replaced by a photograph of a woman's face looking straight at you. After another second, the photo is replaced with the same face gazing either to the left or right of the screen. A half second later, a second photo appears on the right or left of the woman's face. This second photo is of a household item (saucepan, teapot, screwdriver, axe), and it is your job to indicate by button press, and as quickly as possible, whether the item belongs in a *garage* or *kitchen*. Sometimes the item appears on the same side the woman is looking and sometimes on the opposite side. Further, each of the 36 household items is presented in four different colors (red, blue, green, and yellow) during the experiment. You'd see multiple repetitions of these different household items, presented on each side of the face, being gazed at or not.

Do you think the speed with which you'd categorize the items would be influenced by the woman's gaze? Actually, it doesn't matter in this experiment. What matters are your responses on a last surprise task. After you went though the categorization trials, you would be asked to actually give your opinion of the household items on a nine-point scale. As each item appeared, you'd give it a high score if you liked it and a low score if you didn't. The items would appear one by one, each presented in the four different colors, all randomly mixed together.

Here's the amazing part. Your opinion of a particular item of a particular color would depend on whether it was gazed at by the woman's face in the first part of the experiment. Unbeknownst to you, one color version of each of the objects was always gazed at during the experiment. The other color versions of the objects were never gazed at. So if, say, the green saucepan was always in the direction of the gaze, you would rate that you liked it more than the blue, red, or yellow saucepan. The woman's gaze would influence how much you liked an object. When asked at the end of the experiment what most influenced your ratings, you might say the color, shape, or usefulness of the object. But you'd never mention the woman's gaze, and would have no awareness of the relationship between her gaze and a particular color version of each object.

This also is true of your opinion of the gazer herself. It turns out that without knowing why, you will find faces that repeatedly gaze at target objects to be more trustworthy than those that gaze away. Subjects in this experiment were completely unaware that some faces always looked at the objects and some didn't. This suggests again that the influences of gaze can work at an unconscious level.

So would you trust Dave Thorsen's face after being deceived repeatedly by his act? He hopes so. But even if you didn't, chances are you'd still fall for his misdirection techniques. Your own gaze would usually follow his. But even if it didn't, chances are your visual *attention* would.

Here's how. You put on an eye-tracking cap and are told you'll watch a video of a magician performing a simple trick. The video begins, and you see a man standing with his hand facing up and positioned at chest height. From this hand, the man tosses a red sponge ball to a position just above his head (and just below the top of the video's frame) and then catches it again as it falls. He performs this simple toss three times, each time following the ball's path with his eyes. But on the third toss, the ball seems to leave the man's hand and magically vanish. Voilà! The trick is a success.

How do you think it was done? If you were a typical subject in this experiment, you'd guess that on the third toss, the ball left the magician's hand and continued up above the top of the video frame where it was caught by an unseen person. Regardless, you'd likely report having seen the ball leave the man's hand on the third toss.

Time to watch the trick again. You watch the same magician perform the same first two tosses, but this time, on the third toss, he does something a bit different. Rather than following the ball with his eyes, he fixates on his *hand* after the "toss." Aha! Now you know how the trick was done. It turns out that on the third toss, the man never actually threw the ball at all. Instead, he performed the tossing motion with his arm, but actually kept the ball palmed in his hand, unseen. (The compressibility of the sponge ball allowed him to hide it in his palm behind his outstretched fingers.) What allows you to see through the trick this time is what he did, or *didn't* do, with his gaze. The trick worked the first time because the man used his gaze to draw your attention into erroneously predicting the upward trajectory of the ball. Once he withheld his gaze tracking, the trick stopped working.

Of course, in the actual experiment, subjects never saw the trick twice. That would, after all, violate the magicians' creed—and for good reason. Instead for one group of subjects the magician performed the trick with gaze misdirection, and for another he did not use misdirection. The subjects in the misdirected group were twice as likely to fall for the trick. Again, the misdirection of gaze worked like magic.

But maybe not in the exact way you'd think. If, as a subject, you were asked where you were looking during the trick, you'd likely say that you followed the movement of the ball the entire time. But the eye-gaze tracking measures would tell a different story. It turns out that regardless of which subject group you were in, you'd actually have spent about equal time fixating on the magician's face and on the ball's path. This probably means that you were using the magician's gaze to predict the path of the ball.

But the most surprising thing revealed by the eye-tracking measures is that while *you* fell for the trick, your *eyes* didn't. Consider the third, phony throw. Even if you saw the magician's eyes follow the illusory ball's upward motion, your eyes wouldn't track up as much as they had for the first two *real* throws. In a sense, your eye movements knew more about the ball's true location than did your conscious experience. This might be related to how your eye movements often react in a more reflexive way to moving objects. Regardless, the trick really seems to work because of how your *attention* is drawn by where the magician looks. And in lieu of drawing the movements of your real eyes, a magician's gaze can draw the attention of your mind's eye.

YOU INADVERTENTLY SYNCHRONIZE YOUR MOVEMENTS WITH OTHERS

TRY THIS. MAKE A LOOSE fist with your right hand and then put it down on some flat surface such as a table, bed, or your lap. Your fist should be sitting up as if you're pounding the table, your pinky should be making contact with it. Your hand should be making a *relaxed* fist so your fingertips lightly touch your palm. Now extend your index finger out while keeping your other fingers in place. Start waving your pinky back and forth (toward you and away from you) at a comfortable pace. You should wave your index finger at its *preferred* tempo: a relaxed tempo you'd be able to maintain for an extended period of time. Now stop, lift your right hand, and put your left hand down on the surface in the same way. Make sure you relax your left hand. This time, extend the *middle* finger of your left hand and start waving it at *its* preferred tempo. Again, it should be a tempo you could sustain indefinitely. If you're like most people, your middle finger's preferred tempo will be a bit slower than your index finger's preferred tempo.

Now let's try both at the same time. Put both fists down on the surface, and make sure they're relaxed. Start waving with the middle finger of your left hand at its preferred tempo. Continue this, but now also wave the index finger of your right hand, but at *its* own preferred tempo. Because they each have different preferred tempos, they should, in principle, be waving at different rates. But are they? Probably not. If you're like most people, it will be very hard to keep your index and middle fingers from synchronizing (entraining) their tempos and movements. You can even try intentionally moving your index finger faster and you'll probably find that it either inadvertently slows down or your middle finger inadvertently speeds up.

This demonstration shows that despite your body being made up of appendages with varying preferred movement tempos, it prefers to synchronize these movements. The most obvious instance is when your arms and legs synchronize as you walk. Your arms' natural swinging tempo is faster than that of your legs. (You can demonstrate this to yourself by standing on one foot and swinging first your arm by itself, and then your leg by itself.) This is mostly a result of your arms being shorter and lighter than your legs. But when you start swinging them together as you walk—click! They automatically synchronize.

What does this have to do with imitation? Well, it turns out that this inadvertent synchronization that occurs within your body also occurs *between* bodies. Your movements tend to synchronize with those of other people.

Imagine yourself in this experiment. You sit at a table on which is propped a small horizontal handlebar. You're told to grip this handlebar with your right hand and then comfortably extend your index finger out. You're then told to close your eyes, and wave your index finger up and down at its most comfortable tempo. You're to do this until you hear a tone, at which point you're to continue waving your finger at its comfortable tempo, but are to open your eyes and look straight ahead. So, you're comfortably waving away and then—beep! You open your eyes, and there's another person sitting across from you also waving an index finger. And despite having different preferred waving tempos, within one or two seconds, your finger waving would synchronize. You'd be waving your fingers up and down together. This synchronized movement would be maintained until you were instructed to again close your eyes. With eyes closed, your finger waving would not be synchronized with the other person's. But once your eyes were open again, you'd be back in sync.

Unintentional synchronization between people has been observed in many other contexts. Depending on the actions performed, however, synchronization isn't always this consistent. This is especially true when individuals are waving larger appendages or objects. Under these conditions, individuals shift in and out of sync when they watch each other.

Yet they unintentionally end up spending much more time in sync than when not seeing each other. For example, if you were asked to sit in a weighted rocking chair and rock at a comfortable tempo, you would unintentionally synchronize with another person you saw rocking. This synchrony wouldn't be consistent, maybe occurring about a third of the time. But this amount of time in sync would be five times greater than if you weren't looking at each other. And, interestingly, you'd be unaware that you were synchronizing, indicating its unintentional nature.

Interperson synchrony can occur through auditory as well as visual means. This has long been known about crickets, who synchronize their chirps rapidly when located within an audible distance from one another. (This is similar to how oriental fireflies visually synchronize their blinking.) A human example of auditory-based synchronization is something you may have experienced as a member of an audience. Sometimes during prolonged and enthusiastic applause, the clapping will go from being random and incoherent to suddenly synchronizing so that everyone is clapping in unison. Depending on the event, this can actually happen multiple times during an ovation (it seems to be particularly common at European opera performances). Certainly, members of an enthusiastic audience will actively try to maintain synchronized clapping once it occurs, but *when* and *how* it begins is almost always inadvertent.

Unintentional synchrony through sound has also been observed in the lab. Imagine this experiment. You're asked to perform a "find the differences in the pictures" puzzle task with a partner. Each of you looks at separate but similar pictures and talks to determine which subtle features are different about them. As you perform the task, you're standing and facing away from your partner, toward a post that holds your picture. Small sensors are attached to your head and waist using Velcro bands. You're told that you and your partner have two minutes to find as many differences between the pictures as you can. You're also told that as you interact, you need to keep your feet in the same position on the floor, but otherwise can move your body freely. You perform the task with your partner four times. After you're done, you will now perform the same task with a new person (actually an experimenter) who will sit to the side, out of your view. As you do this, your first partner performs the task with another experimenter in the same way.

What's this all about? As you performed the task with both your partner and the experimenter, your postural changes were being monitored. The sensors attached to your head and waist allowed the researchers to track how your body inadvertently swayed back and forth in subtle ways. This swaying happens whenever you stand and talk (or just stand and breathe, for that matter). The researchers were monitoring the postural changes of your partner as well. And guess what: as you were interacting with your partner, your postural changes synchronized to his. The details of this synchronization are very complex, as are the movements of postural sway. It's not as if you were literally swaying back and forth together throughout your interaction. But the amount of time during which aspects of your subtle postural movements synchronized to your partner's was substantially greater than when you were not interacting with him (and were instead interacting with an experimenter). Of course, you would have no awareness of this synchronization; it would be completely unconscious and inadvertent. Regardless, it seems that you are literally in sync with the people you talk to.

■   ■   ■   ■   ■

THERE ARE MANY TIMES when we *intentionally* synchronize our movements with each other. Dancing, playing sports, performing in a band, even moving furniture all involve synchronized movements between people. And the research testing intentional interperson synchronization has revealed something truly fascinating.

But before you learn about that, try this. Put your two fists back on the table as they were before. This time, wave with your two index fingers. Start waving them back and forth but with *opposite-phase* movements. While one finger is moving away from your body, the other should be moving toward your body. Now, as you continue this opposite-phase pattern, start to speed up your wagging, making your movements faster and faster. Move as fast as you can while trying to maintain the opposite phase. At some point, you'll be unable to do so. Your opposite-phase motions will give way to in-phase motions: your fingers will be toward your body and away together.

This shift from one type of coordination to another as speed changes is very common in all types of limb coordination. It's similar to when horses change from a walk to a trot to a gallop. In general, it is the case that the opposite-phase and in-phase coordination styles are the two most common. In phase tends to be more stable than opposite phase. (If you start the index finger demonstration with your fingers moving *in phase*, speeding up won't shift your movements into another type of coordination.) In addition, the subtle deviations from these types of coordination depend on how different the limbs are: swinging two index fingers shows more exact opposite-phase movements than swinging a finger and a leg. Further, the *degree* to which these subtle deviations occur depends on the preferred tempos of each of the body parts involved. Because your finger wants to go faster than your leg, when they are coordinated in an opposite way, your finger reaches its peak position just before your leg reaches its bottom position.

All of these properties of coordination have been shown in multiple experiments with human subjects. These experiments usually involve having a subject swing two handheld pendulums of different lengths, weights, and, therefore, preferred tempos. Certainly, the muscle-tendon sense you learned about in Chapter 7 plays a critical role in your inter-limb coordination. But there is something more universal about what happens when you coordinate. It turns out that many of the characteristics of your coordination described in the preceding paragraph are simply predictable from fundamental physical laws. When your limbs coordinate, there's an important way in which they act like simple, synchronizing oscillating systems—such as connected springs, or cuckoo clocks affixed to the same wall, swinging in synchrony. Similar properties have been observed in other types of biological systems. Despite how we like to think of ourselves as sophisticated, complex beings, run by free will, sometimes we're not much more than conduits for physical laws.

Amazing, yes—but wait, there's more. Not only do these characteristic properties occur between the swinging limbs of a person, but they also occur between the coordinated swinging limbs of *two people*. Evidence for this has come from experiments involving pairs of people swinging handheld pendulums. Subjects sit across from each other and attempt to coordinate their swinging movements. They are instructed to coordinate with either in-phase or opposite-phase movements. Sometimes they're also instructed to speed up their swinging as much as they can. Often, the length and weights of their individual pendulums vary.

These experiments have revealed that when two people coordinate, their in-phase movements are more stable than their opposite-phase movements. Also, the differences between the preferred tempos of their pendulums determines how closely coupled their coordination is. And this difference also determines which pendulum leads the movement by just a bit. In other words, they show the exact same characteristic properties of coordination displayed for *intra*person coordination, as well as connected springs and pendulums.

You coordinate with another person like you coordinate your limbs: by taking advantage of some very fundamental principles of physics. And as long as there is perceptual information passed between the components of coordination, whether via a muscle-tendon system or two visual systems, the characteristic physical properties are at play. So even some of your most sophisticated behaviors—social coordination—can be explained by a simple fact. You're part of the physics of the world, and the physics is part of you.

SYNCHRONIZING WITH OTHERS HELPS YOU PERCEIVE AND SOCIALIZE

WHILE COORDINATING WITH ANOTHER person might take advantage of fundamental physical principles, other factors can influence how and when you synchronize. This is especially true of inadvertent synchronization. For example, there is evidence that even without touching, a mother and her infant show more synchronous behaviors than when a mother interacts with someone else's infant. Also, you're much more likely to inadvertently synchronize your index finger waving with a computer-animated finger if it moves in a natural versus overly smooth (sinusoidal) way. This result is reminiscent of the findings showing activation of mirror system brain areas from watching human, but not robotic, movements.

And like your inadvertent imitation of mannerisms and movements, there seems to be a perceptual advantage to being in sync with others. Imagine being in this experiment. You sit, and are asked to put your right forearm and wrist facedown on a table. Your job is simply to move your right hand up and down so that it is tapping the table in time with a heard metronome. The metronome runs at a comfortable speed, and it is very easy for you to keep up with your hand tapping. You're then told that your tapping movements will be monitored to determine to what degree hearing simple distracter words disrupts your ability to tap. The distracter words are spoken live, by a female experimenter who sits just across from you at the table. You're asked to try to ignore these distracter words but to look at the experimenter's face as you tap.

You hear the metronome begin and tap along. About once every three seconds, she utters a distracter word: *stamp*, *spoon*, *shirt*. You ignore the words as best you can, and listen instead to the metronome. You also notice that besides uttering the words, the experimenter happens to be moving her own hand up and down to the beat of the metronome, along with yours. Then after 60 seconds, the metronome stops and you're told that this part of the experiment is over.

How do you think you'd do with your tapping? It doesn't really matter. What *does* matter is how you'd perform on two additional tasks. After the tapping part is over, the experimenter leaves, and a new, male experimenter enters the room. He asks you to write down all the distracter words you can remember the first experimenter uttering. After giving you about five minutes for this task, he takes out three very similar photographs. He tells you that one of them is an actual photo of the *first* experimenter's face. The other two photos are of her face, but have been subtly morphed (20 and 30 percent) with another similarly aged woman's face. Your job now is to pick out which of the three photos is actually of the first experimenter.

How would you do on these tasks? Well, you're in luck. You were in the group of subjects whose tapping was synchronously mimicked by the first experimenter's *own* tapping. Another group of subjects performed the exact same tapping task, but their tapping was *not* mimicked by the experimenter. By virtue of being mimicked in a synchronous manner, you would perform better on both the word recall and face recognition tasks than those subjects. By synchronizing a behavior to yours, the first experimenter facilitated your inadvertent encoding of the surrounding information.

Not only does synchronous behavior facilitate your perception and memory, it can also create a feeling of group affiliation. It's well known that organizations, ranging from armies to church congregations, perform synchronous behaviors. Whether marching, singing, or group dancing, it's been speculated that these forms of synchronous group behavior help induce an allegiance to the organization and a shift to prioritizing the group over oneself. In fact, very recent research supports this idea. Imagine being told that your task as a subject is to take a short stroll with two other subjects around your campus. The only other requirement is that as you stroll, you must walk in step with your partner subjects. The three of you then come back to the lab and take part in a little game. Without going into detail, the game involves your choosing a number that has different associated payoffs depending on what numbers your partners choose. It is designed to test the relative cooperative and self-serving choices of subjects. The game also tests what subjects assume about *each other's* cooperative and self-serving choices.

What type of strategies would you use for this game: more self-serving or more cooperative? It turns out that by virtue of having walked in step with your partner subjects, you'd be more likely to choose numbers indicating a *cooperative* rather than self-serving strategy. In fact, your choices and assumptions of your partners would be more cooperative than subjects who had just strolled the campus *without* walking in step. Similar results in the game task have been found when subjects are asked to synchronize their behavior by singing in unison, versus singing the same song not in unison. Interestingly, not only are subjects more likely to make collective versus self-serving choices after synchronous behavior, they are also more likely to report a feeling of connectedness and trust with their partner subjects.

WHILE MOST OF US don't have the imitation skills of Marilyn Michaels, we are constantly using imitation for social and perceptual purposes. Whether this imitation is covert or overt, whether it is intentionally guided by the likes of Dave Thorsen or inadvertent, we are, at the core, perpetual imitators. And our imitation helps couple us to each other.

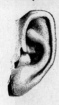

Part VI

Multisensory Perception

Point-light videos demonstrate how our faces can be recognized from only their movements. The technique also allows researchers to determine the important information used for lipreading.

CHAPTER 10

See What I'm Saying

WHEN SUE THOMAS STARTED WORKING FOR THE FBI, SHE DIDN'T think she'd be doing surveillance of organized crime. She was initially hired, along with eight other deaf individuals, to classify fingerprints. The FBI reasoned that the deaf might be well suited for the attention-demanding work of fingerprint analysis, based on the fact that the deaf wouldn't be distracted by the usual office noises. And most of the new employees did take well to the job. But not Thomas. After a few days, she found counting fingerprint loops unbearably monotonous and conveyed this to her supervisor.

Soon after, she was asked to report to the bureau's front office. She assumed that she was being fired. Instead, she was confronted with seven of the FBI's top officers, who were interested in discussing, of all things, her lipreading. A number of her supervisors noticed that during their interactions with Thomas, she seemed particularly skilled at lipreading. Now, they were interested in whether she could also lip-read from videos. She told them that she could, especially if the video showed a clear view of the face. This is what the officers were hoping to hear.

Later that day, Thomas was brought into a room containing a state-of-the-art video projection system. She was told that a surveillance team had recorded video of a suspected illegal gambling transaction involving organized crime. The video didn't contain an audio track, so they needed Thomas to transcribe the dialogue from lipreading. Using the video projection system to enlarge the faces and slow the action, Thomas was able to successfully complete the transcription. And with this, she became the FBI's first professional lip-reader.

Thomas worked with the FBI for about three years. During this time, she lip-read from videotape as well as during on-site surveillance missions. For the latter, she would go to locations such as airport terminals and sit at some distance from the suspects. She would lip-read them as they interacted and then convey the dialogue to her FBI colleagues also positioned nearby. Her cases often involved high-profile crimes including espionage. Thomas's skills provided the FBI a significant advantage to be able to perform accurate surveillance without needing to *hear* what was occurring.

This story is conveyed in Thomas's engaging memoir *Silent Night*, as well as in the syndicated television show based on her book, *Sue Thomas: FBEye*.

■　■　■　■　■

WHAT MAKES THOMAS SUCH an astounding lip-reader? I ask her when we meet. Thomas, who is in her late fifties and has a warm, cheerful face, responds, "I was never actually *taught* to lip-read. But at a young age, I was given seven years of intensive speech therapy to make my own speech clearer. This therapy involved lots of time in front of a mirror watching and correcting my speech. This helped me speak, and also probably helped me become a good lip-reader. But I never did have a course in 'Lipreading 101.'" Thomas smiles.

And this intensive speech therapy has truly paid off. Thomas's speech is *very* clear and has allowed her to become a professional speaker since leaving the FBI. And her lipreading truly is amazing. During our 40-minute discussion, she only asks me to repeat myself twice: both times, she claims, because I was smiling too broadly as I talked.

Thomas lost her hearing to an unknown illness when she was 18 months old. Her parents felt strongly that she be integrated with the hearing community and enrolled her in an intensive speech therapy program. "The word repetition was constant. Not just at the school, but at home. My [hearing] mother would stand in front of the bathroom mirror with me for hours, and continue what my teacher was doing at school. In fact at that time, my mother also became a very good lip-reader. Whenever we were at a party and she sensed that I was feeling left out, she would talk to me from across the room by mouthing her words. I would do the same in response, and we could have conversations this way—by lipreading each other."

Despite the challenges she faced as a deaf child, Thomas became a state ice-skating champion at the age of seven. She also learned to play classical piano by feeling the vibrations through the keys. Thomas went on to college where she graduated with a degree in political science and international affairs. After working as a counselor for parents of deaf children, she was hired by the FBI.

I ask Thomas whether she used any different lipreading strategies when she was doing her surveillance work. She answers, "Not really. However, I did learn to use the 'whole context' at the FBI. Usually, people's body language and conversations match each other. But when people are being secretive, that sometimes doesn't happen. By being aware of this secretive context, I could see past the body language and concentrate on the dialogue."

I ask whether lipreading professionally allows her to now see more details of conversation, and from further away. She replies: "Well, I knew pretty much in high school that I was able to understand people's conversations from a distance. In fact, when I realized this, I made a vow to never say what I saw, because it's an invasion of privacy. And frankly, there have been times when I've seen people say things I wished I hadn't seen."

Later in our conversation, I ask Thomas if she knows anyone who's tried the commercially available lipreading training programs. In response, she laughs and says, "I wasn't aware that such things existed. I'd be surprised if they worked all that well. But I do think that good lipreading is a skill any deaf person can learn. The trick is learning to carefully say the words correctly yourself. In fact, when parents of deaf children ask me if I have any advice, this is what I tell them: I say that even if their children exclusively use sign language, and never, ever speak in public, they should still have speech therapy. Why? Because speech therapy enhances lipreading skills. I'm not sure you can just teach someone to read lips. But by teaching someone to carefully *pronounce* a word, you've taught that person how to read the lips for that word."

■　■　■　■　■

BEYOND SUE THOMAS'S INTENSIVE speech therapy, there are other reasons for her superb lipreading skill. The fact that she became deaf at an early age probably played a role. While lipreading skill varies widely in both hearing-impaired and hearing populations, research shows that individuals with early-onset deafness are generally better at the skill. This is especially true of individuals whose deafness is profound and, like for Thomas, began before the bulk of their language acquisition (before three years old). It's thought that relying on lipreading to acquire language puts an individual at a strong advantage for superior lipreading skill throughout life. Recent research shows that early-onset deaf individuals are generally able to lip-read, from a silent face, more than twice as many words and sentences than hearing individuals. This difference is likely greater for lipreading in conversational settings and from familiar talkers, which, Sue Thomas feels, helps her significantly.

There are other characteristics that probably make Thomas a good lip-reader. Not only did she start lipreading at an early age, but she also has a lifetime of experience using it as her chief means of communicating. Research shows a link between superior lipreading skill and its consistent use in adulthood. The research also suggests that being a woman provides her an advantage. Finally, Thomas is a very bright person, and high cognitive abilities have often been shown to play a role in superior lipreading skill. Measures such as short-term memory span and the speed with which one can recognize *written* words correlate with lipreading ability.

At the same time, some of these attributes may be a *result* of early-onset deafness. In ways similar to how brains change with blindness, brains change with deafness. There is evidence that both the seeing and hearing areas of the brain reorganize in early-deaf individuals. For these individuals, the visual brain areas reorganize to allow for greater reactivity to visual motion. This enhanced sensitivity is more pronounced for tasks requiring greater visual attention, as well as for images seen in the periphery. Early-deaf individuals' visual brain areas also shift so that the *left* side of their brain (for right-handed individuals) is more reactive to visual motion than their right; a reversal from that of most hearing individuals. This may also provide for greater sensitivity to visual changes and motion.

There's also evidence for *cross-modal plasticity* in the deaf brain. Many regions normally dedicated to hearing (auditory cortex) get used for seeing. These regions have been shown to respond to visible sign language in the brains of deaf individuals, as well as other types of hand and arm gestures. This has also been shown even for simple moving-dot patterns.

This plasticity in both the visual and auditory brain areas supports some enhanced perceptual skills in early-deaf individuals (once other confounding factors are considered). The early-deaf are faster and more accurate than hearing individuals at detecting the onset and direction of simple motions. This is especially true with motion seen in the peripheral field of view and when the task requires greater attention (when other images must be ignored). In fact, with regard to the field of view, visual attention works in an *opposite* way for deaf and hearing individuals. While it is easier for hearing individuals to be distracted by images appearing in the central field of view, deaf individuals are more distracted by images appearing in the *periphery*. This makes sense. For those of us who hear, we can use our ears to monitor locations where we're not currently looking. If something surprising appears in the periphery, we can hear it. For the deaf, peripheral surprises must be monitored by vision. It makes sense, then, for cross-modal plasticity to enhance visual attention to the periphery of a deaf individual's visual field.

In a way analogous to the blind, individuals with early deafness show a cross-modal brain and perceptual plasticity allowing for compensatory skills.

But what about those of us whose hearing impairment isn't early onset, or even that profound? There are many of us, after all. Most recent estimates show that over *30 percent* of U.S. adults have some form of hearing loss. Sixteen percent have the type of loss that directly hinders the ability to perceive speech. By the time Caucasian men are in their forties, nearly 60 percent of them will have some hearing loss (women and individuals of other ethnicities lose their hearing less quickly). Most troubling are results showing that the prevalence of hearing loss in *young* adults, ages 20 to 29, has been growing steadily for the last eight years.

It's a good thing you have a way to partially compensate for hearing loss. While it's unlikely you have the lipreading skills of Sue Thomas, you do read lips, and you do it all the time.

You Read Lips

Try this. Go to your TV and turn on a station that's showing a close-up image of a talking face. You'll want a show that provides a consistent image of a talking face, so C-SPAN or a news show would be a good choice. Now, close your eyes and turn down the volume on the TV until it is barely audible. It should be low enough so that you can hear that someone is speaking but *not* be able to understand what they're saying. It may take a minute for you to find the right level. Once you find the right level, and you're sure that you can't quite make out what's being said, slowly open your eyes and look at the TV. As you look at the speaker's mouth, you should now be able to make out much of what's being said. You're lip-reading. (If this demonstration doesn't work that well, look for another talker on the TV: Some people are definitely easier to lip-read than others.)

When you're listening to someone talk in a noisy restaurant, you're lip-reading. When you're in a quieter environment and listening to someone who speaks with an accent, you're lip-reading. Even if it's quiet and the talker doesn't have an accent, if the discussion is complicated, you're lip-reading. You lip-read when you're learning a new language, and you were lip-reading when you learned your first language.

Research shows that you lip-read in all of these contexts. This research suggests that whenever you're able to see a talker's face, you watch and use the speech movements, and have been doing so all your life. You lip-read to enhance the speech you hear. And while you're not able to lip-read as much in silence as Sue Thomas, you use the same facial information, same processes, and much of the same neurophysiology she does.

Of course the lipreading you perform is largely unconscious. It is truly one of your implicit perceptual skills. But your lipreading is also automatic, and in some cases *overwhelming*. The next time you have two friends nearby, try this. Have one friend stand just behind the other so that they're both facing in the same direction. Now face the friend who's standing in front. You should be able to see the front friend's face, but not the face of the friend who's in back. Now have the friend in back repeatedly utter the syllable *ba* at a comfortable rate while the friend in front *silently* mouths the syllable *va*. Once they're in sync, watch the face of your friend in front, but concentrate on what you *hear*. You should be hearing a pretty clear *va* syllable despite your ears only receiving the *ba* syllable uttered by your friend in back. What you see is overriding what you hear. You are experiencing the power of lipreading. (If you don't have two friends nearby, there are many terrific video examples on the Internet.)

This phenomenon is known as the McGurk effect, named for the lead scientist who first discovered it. The effect can be quite strong. In our own laboratory, we find that a visible *va* overrides an audible *ba 98 percent of the time*. And the effect can occur in many different ways. If your friends are still game, you can have your friend in back repeatedly utter *ba*, while your friend in front silently mouths *ga*. You should now perceive *da*. In this case, you're perceptually compromising between what you're seeing and hearing. To understand this compromise, try slowly articulating *ga, da, ba* yourself. You'll notice that in terms of where you're obstructing your mouth, *da* is right in-between *ga* and *ba*.

The McGurk effect is also usually quite durable. Ask your friend who's in back to now stand behind *you*. Have your friends produce the visible *va*, heard *ba* described earlier. Despite the fact that the face and voice are clearly coming from different positions, the effect should still work. Next, if your two friends are the same gender, have the audible friend recite the syllables, pretending to be the opposite gender. It shouldn't matter: knowing that the voice you hear is different from the face you see—even if of different genders—doesn't eliminate the McGurk illusion.

The McGurk effect works with speech perceivers of every language background for which it has been tested (although the specific *types* of speech integrations depend on the language). Even for scientists who've been working with the effect on and off for 25 years, have created McGurk video stimuli, and whose own faces and voices have been *used* for the stimuli, the effect still works. I can vouch for that myself.

The McGurk effect is one of the most referenced phenomena in speech perception research. The effect demonstrates that we all lip-read and use visual speech information. It shows that our use of this information is immediate, automatic, and, to a large degree, unconscious. It shows that despite our intuitions, speech isn't just something we hear. More generally, the effect, along with the thousands of studies it's helped motivate, has shown that our brain is designed around multisensory input. In an important way, your brain doesn't care through which sense it gets its perceptual information. And for some functions, it might not even know. But more about this later.

The McGurk effect has also been used as a tool to understand the type of facial information we require for *visual speech perception*. It's a useful tool because it doesn't require subjects to explicitly lip-read. Instead, subjects are simply asked to perform something very natural: just report the speech they hear. Still, the effect persists. The effect works whether you're looking at a face from the front or the side, or whether the face is extra large or extra small. It works with faces that can barely be seen because of a very low contrast video image, or if the face is shown upside down. Interestingly, the effect is sometimes reduced with upside-down faces, for reasons you may remember from Chapter 8. Our own lab has shown that the effect works if the face is reduced to nothing but moving white lights against a black background: the point-light technique you learned about in Chapter

8. Interestingly, when we ran this experiment, some subjects never realized they were seeing a face. Yet even those subjects showed the McGurk effect. It seems you can lip-read without even knowing that you're looking at a face.

You Lip-Read from All Over the Face

OTHER RESEARCH HAS REVEALED the different types of speech information that you can perceive from a face. Try the TV demonstration again and concentrate on the face of the talker. What pieces of speech can you identify? Perhaps you can recognize when the speaker utters a *b* or *v*, or other consonants using the lips. Many vowels might be visible because they also use the lips. The start and stop of words and sentences might also be recognizable, as well as the emphasis used by the talker. It seems that as long as a speech component uses the lips, then you might be able to use it as visual speech information. It is the lips, after all, that you use when *lip*reading, correct?

Incorrect. It turns out that you're looking at a lot more than the lips when you lip-read. Research in our lab and others shows that without also seeing what the tongue and teeth are doing, you're not getting much useful information at all. This research shows that you also use information from the cheeks and jaw. As you lip-read, you actually use information from across the face. This helps you detect subtle aspects of articulation that you wouldn't think would be visible.

While a talker's tongue-tip motions might be relatively visible by looking at the mouth, you would think that the back of the tongue would be pretty much invisible. But not so. Careful analyses of facial motions have revealed that tongue-back position *is* visible. But not through looking at a talker's mouth. Instead, by looking at subtle motions of a talker's *jaw*, you can tell something about the position of the tongue back. Knowing the position of the tongue back is important for distinguishing many vowels and consonants.

Even more amazing is that visible facial motions can provide information about *air*. Specifically, there is information about the changing amount of air inside a talker's mouth: air changes that can distinguish, for example, those voiced from unvoiced consonants (*ba* versus *pa*) you learned about in Chapter 9. It turns out that the very quick air pressure changes occurring in a talker's mouth are visible in subtle movements of the *cheeks*. You may be able to notice these cheek changes yourself by looking in a mirror (or placing your hand on your cheeks) as you slowly and exaggeratedly produce a *pa* versus *ba*. So, the visible speech information spread across a talker's face allows you to perceive the "hidden" speech components of tongue-back position and intramouth air pressure. When it comes to lipreading, it's as if you have x-ray vision.

But not only are these hidden components of vowels and consonants visible, the more *musical* aspects of speech are visible. When you produce sentences, and even words, you produce them with a change in pitch. This is most obvious when listening to yourself produce the same sentence as a question and then as an answer. This is true? This is true. But you are actually varying sentence pitch—or *prosody*—up and down *throughout* a sentence as you stress words and syllables. This is a fact you can readily notice by slowly reading this sentence out loud. Now, you largely control sentence pitch with the tension in your vocal chords: you pull them tighter to raise pitch as you would at the end of a question. But your vocal chords are way down in your throat so you'd think that speech pitch changes wouldn't be something you could see from lipreading. You'd be wrong.

Recent research is showing that you can lip-read pitch changes— but not how you'd think. To lip-read the speech pitch change occurring in a talker's throat, it's best to look at the *forehead*. In one experiment, a talker was videotaped producing sentences. Acoustic and visual analyses of these recordings revealed a close correspondence between the changing pitch of the sentences and the talker's head movements. The pattern of head motion corresponding to pitch was complex, but as the pitch went down during the sentence, the forehead also subtly dipped.

So visible head nodding is related to pitch. But does it help you perceive speech? It seems it does. Using the head and face movement information measured from the videotapes, the same researchers computer animated a talker speaking the same sentences. One version of the animated talker included both the face and head movements of the original talker, while another just contained the facial movements, with the head still. When subjects watched these animations together with auditory speech made to sound noisy, they were able to perceive significantly more of what was being said when the talker's head moved in a natural way. Watching head movements helps you lip-read better.

Related research shows that you can distinguish sentences from statements by watching head and even *eyebrow* movements. These experiments also showed that you can use similar information to help determine the end of a sentence. Other research shows that you can distinguish speech "tones" (used in Mandarin, and other Asian languages) also from watching head movements.

So given that there's visible speech information spread across a talker's face, where do you look? At the eyes, of course: it's the polite thing to do—at least in a majority of cultures. The eyes themselves are not a particularly rich source of visible speech information. As you've learned though, there's a good deal of other information to be gained from watching someone's gaze. But even when asked to explicitly perceive spoken words from a video of an audible talker's face, research shows that you would spend much of your time looking at the eyes. This may be partly a result of your natural gaze behavior with faces. Also in most cases, your more peripheral vision is sufficient to perceive the visible speech around the mouth and other speech-relevant areas. However, when background noise is added to heard speech, things begin to change. You would start to look more toward the lower half of the face, centering around about the nose. And you would look more intently at this area, ending the gaze shifts typically performed when speech is more audible. In fact, this is something even expert lip-readers will do under more difficult conditions. Sue Thomas readily admits that when she is tired, or there are distractions nearby, she will look at the lower half of a talker's face. And this makes sense. The mouth is probably a better source for lipread information than most other *single* positions on the face.

Still, the research shows that in most situations, you *lip*-read from a talker's tongue, teeth, eyes, mouth, jaw, cheeks, head, and even eyebrows.

This has led some researchers to switch from using the term *lipreading* to *speechreading*. These days, the two terms are used equally often. And as a researcher in the field, I've used both the "lip-" and "speech-" prefixes, and find them equally acceptable.

It's actually the "reading" part of the term I find misleading. After all, visual speech perception is much less like reading than it is like *hearing* speech. Reading involves using an arbitrary code (letters). Heard and lip-read speech does not. Reading also involves *formal* training to use. Certainly this isn't necessary for heard speech. And while it may be possible to improve your lipreading a bit with formal training, neither you nor Sue Thomas has had explicit lipreading training, and you've both lipread your whole lives. Finally, read text can't integrate with heard speech like lip-read speech can. You can't induce a real McGurk effect with *read* syllables (from a computer screen). In fact, you've learned that read text can't even integrate with heard speech as well as *felt* speech can: the felt speech that Rick Joy, and even you, can use. Your speech-perceiving brain is on the hunt for articulation and is willing to take it—and use it similarly—regardless of whether the articulation is heard, seen, or felt. The point is that you, and Sue Thomas, don't really *read* lips. You both *perceive* speech articulation whether it's heard or seen.

One other aspect of lipreading makes it more like heard speech than reading: you probably started to lip-read as soon as you could see faces.

You've Always Lip-Read

"Whoops, we've lost another one." This is not something I expected to hear during an experiment testing a five-month old baby.

I turned to look at the video monitor and was relieved to see that the infant was fine; he had just spit up over himself and his mom, whose lap he was sitting on. "This type of thing happens a lot. It's one of the things that can make infant research very slow." My collaborator and old friend Mark Schmuckler was describing how things typically go with his infant subjects. "Some of them spit up, others fall asleep, many of them just get fussy." I replied, "Wow. That stuff rarely happens with the freshman subjects in our lab—well, except for the fussy."

IT WAS 1996, and I was visiting Schmuckler's lab to help out on our collaborative project. We were interested in whether five-month-olds display the McGurk effect. My lab prepared the audiovideo stimuli; Schmuckler's lab tested the babies. Of course the challenge, besides the spit up, was how to get infants to tell us what they were perceiving: whether *va* or *ba*. Fortunately, perceptual development research has established many ways for infants to tell us a great deal of what they're experiencing.

For our experiments we made use of a fact that every parent knows well: infants get bored with old things and interested in new things. They get especially bored if they see and hear the same thing over and over again. We reasoned that by showing infants different strings of audiovisual syllables, we could monitor their boredom to determine when they perceived something old or new.

To get the infants bored of audiovisual speech, we showed them a face uttering the same syllable, *va*, over and over again. We did this by videotaping a male talker saying *va*, and then repeating this clip so that the infants saw the same loop of videotape every two seconds. Bad idea.

Schmuckler told me the story when I first arrived at his lab. "You guys edited the video clip so that as it looped, the talker's mouth repeated *va*, but his eyes *never changed*. The talker looked straight ahead at the infant, without blinking. It made him look pretty scary. Lots of infants cried. *Not* a good thing if you want parents to bring their kids back for more experiments." By the time I was visiting, Schmuckler had wisely placed cardboard over the top half of the video monitor so the infants couldn't see the talker's spooky stare.

■ ■ ■ ■ ■

AFTER THE LAB ASSISTANTS helped the first mom clean the spit up, a new infant and mother were brought in. The infant was a becomingly chubby five-month-old boy with black hair and blue overalls. He sat on his mother's lap and happily played with a rattle while she was fitted with headphones. It was important that the mom not know when the speech sounds changed, to ensure that she couldn't unconsciously distract her baby when this happened. The infant, on the other hand, was able to hear the speech syllables through a loudspeaker positioned under the video monitor.

The mom took away the rattle, and the videotape began. The infant looked up and stared intently at the monitor as the face repeated *va-va-va-va*. As we watched the infant via a hidden camera, I asked Schmuckler: "How do you know when he gets bored?" (I had never been involved in an infant experiment before.) He replied, "It's pretty obvious with babies. You'll see." And he was right. After about 25 seconds of seeing and hearing *va*'s, the baby's head and entire upper torso flopped down to the right. Schmuckler smiled, "He's bored."

Then the baby was presented with our critical video clip. The baby would again be seeing repeated *va*'s but *hearing* synchronized repeated *ba*'s. We reasoned that if babies are like adults, they will perceive this combination as *va* and get bored more quickly because they had just perceived 25 seconds of *va*'s in the first set. If, instead, they didn't integrate the auditory *ba* and visual *va*, and perceived *ba*, they would find this novel and gaze longer. (This description is simplified for explanatory purposes.)

The tape began, and we watched the infant. At first, he stared intently at the video monitor and we held our breath. And then, after just 11 seconds—flop—his head and upper torso fell to the side. He got bored much more quickly.

This was the pattern for most of the babies in the experiment. It may have meant that like adults, the babies were truly perceiving *va* when presented the visible *va* with a heard *ba*. But we had many more things to test before we could make that conclusion. We needed to make sure that the babies weren't just getting bored quicker to the second video clip simply because it *was* the second clip. To ensure this, we tested other infants with the same audiovisual *va* first, but followed it with an auditory *da* paired with a visual *va*—which adults perceive as *da*. The infants did *not* get bored with this stimulus.

Over the course of the following weeks, we tested many other possible explanations for our original findings. None of them was supported. Our original explanation was right: The infants seemed to have perceived the audio *ba* + video *va* as a *va*, and that is why they got bored so quickly after perceiving 25 seconds of audiovisual *va*. The babies fell for the McGurk effect.

Since our experiment, other labs have replicated these effects with other audiovideo combinations (*ba* + *ga* = *da*) and slightly different methods. Another study used brain imaging to show different patterns of brain responses in infants depending on whether they were presented fusing syllable combinations (audiovisual *ba*, audiovisual *ga*, audio *ba* + visual *ga*) versus nonfusing combinations (audio *ga* + visual *ba*). The general conclusion is that by the time infants are five months old, they generally integrate audiovisual speech the same way adults do. You were integrating audiovisual speech before you could talk.

But what about before five months? It's difficult to use the bored-baby methods with infants much younger than this. But research using other methods indicates that you were at least sensitive to audiovisual correspondences at a younger age. At four months old, you were aware of correspondences between vowel sounds and their lip shapes on a face. At three months, you could detect timing asynchronies in audio and visual speech. At one month old, you would get distressed if your mother's voice came from a different location than her face. At that time, you would also suck your pacifier more rapidly to audiovisually congruent than incongruent syllables. And as you learned in the last chapter, at *one day* old, you imitated vowels with your little mouth whether those vowels were heard or seen. Thus, although it hasn't been definitively established whether you were *integrating* audio and visual speech before five months, you were certainly visually perceiving speech gestures from faces before then.

This turned out to be very useful for you. In fact, lipreading helped you acquire language. One recent experiment demonstrated that infants can use visual speech to help them hear subtle differences between speech syllables. Moreover, the initial presence of visible speech allowed them to later categorize the syllables by sound, even after the visible speech was removed. This finding is consistent with what is known about language acquisition in blind infants. Many blind infants acquire language a bit later than sighted children. Interestingly, the *types* of delays observed in the blind are often predictable from the fact that they can't see faces talking. Blind toddlers often have trouble distinguishing speech sounds that are audibly very similar, but visually very distinct. For example, *m* and *n* are pretty similar in how they sound. You may have noticed this when on the telephone: "*m* as in monkey; *n* as in Nancy." When you were an infant, you learned to *hear* this distinction by looking at faces. Because faces produce *m* and *n* in very distinct ways, this helped you attend to the subtle differences in how they sound. But blind infants don't have access to this visual speech information and are consequently delayed in being able to hear the subtle *m* versus *n* distinction.

Your infant lipreading skills went beyond integrating and detecting correspondences in audiovisual speech. By the time you were four months, you were able to visually discriminate between languages (English versus French) just by watching a person's face. This is a skill you've pretty much maintained through adulthood.

Interestingly, however, there was one time in your childhood when you weren't using visual speech that well. It was when you were learning to read, probably between the ages of five and eight. It's unclear exactly why this happens. One possibility is that the substantial reorganization needed for you to map letters onto sounds borrowed resources you usually used for perceiving speech from faces. Regardless, your visual speech skills reappeared by your adolescence and have likely only improved since then.

YOUR BRAIN HEARS THE SPEECH YOU SEE

WHEN YOU EXPERIENCED THE McGurk effect earlier, you very likely thought you were *hearing* the speech presented visually (*va*, for example). This is an aspect of the effect we've observed in the laboratory again and again. Despite being asked to only report what they *hear*, subjects can't help but be influenced by what they see. Scientists now have a sense of why this occurs. It turns out that when you are presented audiovisual stimuli, response patterns in your *auditory* brain areas (auditory cortex) change as the *visual* stimuli change. Conversely, keeping visual components the same (*va*) while changing auditory components (*va* to *ba*) fails to induce changes in your auditory brain, despite the significant acoustic change (recall that you "hear" *va* in both cases). In the McGurk effect, your auditory brain uses visual speech as if it's heard.

But even when there is *no* sound present, speech or otherwise, your auditory brain responds to lip-read speech as if it's sound. And this response occurs even though you've had no formal lip-reading experience. In the original demonstration of this phenomenon, hearing subjects lay in a scanner as they lip-read from a face silently articulating the numbers one through nine. As they identified the lip-read numbers (a relatively easy thing to do), their auditory brains showed activity akin to when subjects literally *heard* the numbers spoken. Beyond showing how your brain readily uses lip-read speech, these findings are profound. They are some of the first to clearly show that visual information on its own can activate the human auditory brain. The findings helped initiate a rethinking of perceptual integration in general and the multimodal nature of the brain. But more on this later.

Since these original findings, other research has found just how fundamental these influences of lip-read speech on the auditory brain truly are. For example, there is now evidence that the visual influences occur as quickly as those of auditory stimuli, and affect the *specific groups of neurons* in the auditory brain that also respond to auditory speech. Other research has shown that lip-read speech can affect a brain area known to be involved in some of the earliest, peripheral stages of auditory processing (auditory brainstem).

All of this neurophysiological research is consistent with what's been observed back in the perception lab. McGurk effects have been used to induce changes at the earliest stages of speech recognition, before your brain can recognize syllables or even speech sounds (phonemes). Recall that one of the ways you perceptually distinguish the consonants *p* from *b* is in detecting whether there's a delay between when a talker's lips separate and the vocal chords start vibrating. For *b*, there is no delay, and for *p* the delay is usually less than one-twentieth of a second. It turns out that your brain's interpretation of the length of this delay can be influenced by the consonant you lip-read. Specifically, if you were presented an ambiguous consonant sounding somewhere between *b* and *p*, being simultaneously presented a visible mouth uttering a quick *p* would make you hear the consonant as an *unambiguous p*. If, on the other hand, you heard this same ambiguous consonant paired with the sight of a mouth uttering a slower *p*, you would hear an *unambiguous b*. This means that visual speech can shift the perceived lip release/vocal chord vibration delay by one-twentieth of a second. Visual speech can influence what you hear on the most microscopic level.

■ ■ ■ ■ ■

NOT ONLY CAN LIPREADING influence your auditory brain, but it can also influence the brain regions you use to *produce* speech. Recall from Chapter 9 that hearing specific speech sounds can induce neurophysiological and muscular reactions associated with *producing* those sounds. This covert imitation was thought to facilitate your skills at perceiving speech.

It turns out that lip-read speech can do the same thing. In the context of the McGurk effect, the visible component can induce activity in the speech production areas of the brain (premotor cortex, Broca's region) associated with the *fused* perception. Silently lipreading has also been shown to induce activity in speech production areas. Interestingly, activation of motor areas can be induced by visible speech composed of point-light displays.

Lipreading can also induce activity in corresponding muscles with the help of that good old TMS pulse (a very quick pulse of transcranial magnetic stimulation). If you were to have your facial motor brain area pulsed while watching lips articulate vowels, you would show corresponding activity in your own lip muscles. This covert imitation in your lips would not occur when watching eye movements unrelated to speech. And your covert speech imitations would, themselves, be subject to the McGurk effect. This research used the TMS pulse technique as well as recordings from lip muscle. Subjects were shown a number of different visual and audiovisual stimuli. These included a visible mouth articulating *ba*; the mouth articulating *ta*; the mouth articulating *ba* seen together with an audible *ba*; and a McGurk-type syllable composed of a visible *ta* with an audible *ba*. This last combination is almost always perceived as a *da* by subjects. Results showed that only the stimuli *perceived* as *ba* induced motor activity in the lips. Thus, while the visible *ba* induced lip muscle activity, the visible *ta* did not. Unsurprisingly, the audiovisual *ba* also produced lip muscle activity. But what about the McGurk syllable: visible *ta* + audible *ba*? It turns out that this combination did *not* induce lip muscle activity despite the presence of the audible *ba*. The information you get from lipreading can override the speech you hear to the degree that it influences your covert imitation.

And lipreading can inadvertently influence your *overt* imitation as well. Imagine being in this experiment, conducted in our lab. You sit in front of a video monitor and are told that your task is to lip-read words. Don't worry, we'll give you hints about what the words are. On each trial, you see two words printed on the video screen, for example: *tennis table*. Immediately after, you see a talker's face silently uttering one of the words: *tennis*. Your job is to identify the word you lip-read by saying it out loud, clearly and quickly.

You would probably be quite good at lipreading the words in this choice context: our untrained subjects recognize about 95 percent of the words. But more important is the fact that despite no instructions to do so, you'd inadvertently imitate the speaking style of the talker you lip-read. All of your spoken word responses would have been recorded. And your word responses would sound significantly more like the talker's words than if you had just read the words out loud. These results suggest that you lip-read certain details well enough to unintentionally incorporate a talker's manner of speaking into your own. Put differently, the results show that you can inadvertently sound like a talker whose voice you have never heard.

Clearly, these results are related to the effects discussed in the previous chapter. Like heard speech, you have a tendency to automatically mimic the speech you *see*. And like for heard speech, you inadvertently imitate the *fine-grain* aspects of lip-read speech. In another experiment conducted in our lab, we tested whether lip-read speech could induce subjects to imitate the delay between when the lips release and the vocal chords start vibrating. Recall from Chapter 9 that different talkers have slightly different delay times ranging from about one-twentieth to one-tenth of a second. And recall that previous experiments showed that subjects partially imitate this delay based on the speech they hear. We've learned in our lab that subjects also unintentionally imitate a talker's specific vocal chord delay based on how *visible* lip releases are timed relative to heard speech. It seems that you're able to lip-read subtle lip timings well enough so it inadvertently influences your own lip–vocal chord timing in very nuanced ways.

You Can Match Voices to Faces

"I knew right away that John Travolta would make a great puppy," Chris Williams tells me. Williams, director of the Disney/Pixar animated film *Bolt*, is describing how his team selected actors to voice the characters of his film. "It's true that Travolta is now known for playing real tough characters—killers and thugs. But the reason he's so successful with those parts is that he, himself, is such a sweet person. He's a really warm and gentle guy and that comes through in all the characters he plays. So when he's playing tough, you can see a good person underneath, and that makes for an interesting complexity. Now, the character 'Bolt' is a small dog who *thinks* he's a tough guy who can defeat fleets of helicopters and tanks. But we, as the audience, can see what he really is—a sweet, harmless puppy-dog. That's why John was so perfect for the character."

But tremendous effort goes into choosing the right voice for an animated character, especially for big-budget films like *Bolt*. "First we brainstorm about the type of actor that seems similar to the character we want voiced. We also consider actors who we think might be able to *become* the character, even if they haven't played similar characters in the past. We narrow that list down to 10 to 20 actors for each character. We then have our sound editor pull together audio clips from each of those actors' television and movie work. He puts these on a big loop so we can hear an actor say different types of lines, and hear this series of lines over and over again. As we listen to each actor we look at rough sketches of the character on a computer monitor. We all then try to imagine that character sounding like the voice we're hearing. It's actually pretty funny because while the voice is from the same actor, the lines are taken from different shows and don't make sense together. It's why John Lasseter [*Bolt*'s producer and chief creative officer at Disney/Pixar] calls this process 'Non Sequitur Theater.'"

Williams continues, "For some of the actors, we'll know right away that their voice won't work for a character. When heard separately from their face, you may notice a raspy-ness in an actor's voice that will make it wrong for the character. Or maybe the voice will sound the wrong age. Going through this process also makes you notice that some actors, even the best ones, act less with their voice than their face. That doesn't work for animation. Anyway, we'll usually end up with two or three voices which just seem to settle in and sit with a character in a way other voices won't. We'll then debate over which one to use for the character."

Because this process needs to be done for each of a film's characters, it can take many weeks to complete. Once the voice actors are chosen and sign on to the film, they're brought in one at a time to record their lines. This creates another obstacle for directors like Williams. Most modern voice acting for animated features involves recording the actors separately. In fact, this is one of the appealing aspects of voice acting for stars like John Travolta: he can perform his part on his own time, without needing to coordinate his schedule with other actors, or waiting in a trailer for his scene to be set up.

But recording the actors individually poses a technical problem for the director and sound editor of the film. Essentially, they need to create compelling dialogue between actors who never actually talk with each other. For *Bolt*, John Travolta recorded his lines months before Susie Essman (who voiced "Mittens," Bolt's feline sidekick) recorded hers, despite the fact that their characters had multiple dialogues together. Further, actors in a scene perform their lines without hearing recordings of the other actors' performance. Recall from Chapter 9 how people subtly imitate each others' speaking rate, intonation, and manner as they interact. But actors who must produce lines of dialogue without actually hearing one another would be incapable of this imitation. And it seems that when this subtle speech imitation is absent from dialogue, it can sound distractingly unnatural to an audience.

"A good portion of my time working on *Bolt* was spent in the editing room piecing together bits of dialogue so that the characters sounded like they were talking *together*," Williams recalls. "As I was directing, I had the actors perform each of their lines 10 to 30 times with different speeds, inflections—I tried to get a wide spectrum to select from. Once all the dialogue was recorded [and before animation began], we spent many late nights putting the lines together. For every piece of dialogue we'd listen to each of the 30 line readings again, and again, and again to make sure we chose the one that sounded right with the other characters' lines. If you put lines together that don't sound natural, it's *very* jarring and apparent."

This suggests that not only do you subtly imitate the people you talk with, you're also aware when this imitation is missing from the dialogue of others.

All of Chris Williams's hard work paid off. Bolt was a critical and commercial success and was nominated for a 2008 Academy Award for best animated feature. And the voice casting, including that of John Travolta, is compelling. "People will often say to me how the dog Bolt *looks* like John Travolta," Williams says. "But he *doesn't*. Bolt's appearance was pretty much set before John was even cast! And even after we had John's vocal performance, there was no consideration of his appearance in the way Bolt looked. The animators never *saw* John's performance. But people will watch the film, and sometimes— hopefully not too often—recognize they're hearing John Travolta's voice. They'll then maybe see Bolt's character move in a way inspired by John's vocal performance and start seeing John's face in Bolt. It's really interesting."

■　■　■　■　■

WHY IS SUCH EFFORT put into casting voices for animated characters? Part of the reason has to do with one of your implicit perceptual skills. You know how voices match faces. Sometimes the opposite may seem true. Perhaps you've had the experience of listening to someone's voice on the radio for a few years, and then being surprised when you finally saw what that person looked like. The examples where faces don't seem to match voices (Mike Tyson, Julia Childs, Fran Drescher) stand out in our experience. But these are likely the exceptions that prove the rule. For the vast majority of cases, faces *do* match voices, and you are implicitly sensitive to this fact. This has been demonstrated by research. If, for example, you listened to a single word (*cat*) being spoken by a male and were then shown two male faces silently speaking that word, you could correctly choose (at better than chance levels) which face matched the voice. You could also do this with a female voice and two female faces. You could even make these matches if the word you heard (*cat*) was different from the word you saw the faces utter (*dog*), or if you heard one sentence and saw the faces uttering different sentences. And it's not important that you know the *meaning* of the sentences. You could make the voice-face matches for talkers who were speaking a language you didn't know.

How might you match voices to faces? Some of your strategies might be pretty intuitive. You could listen for the voice's rough age, and look for the talker who looks to be closest to that age. You might listen for the pitch or loudness of the voice and then match based on the apparent size of the talkers. And as you'll learn in the next chapter, you could implicitly listen for the attractiveness of the voice and match based on the comparably attractive talker.

But, there must be another type of information you can use to make voice-face matches. Because as it turns out, you can make these matches when seeing faces that don't look like faces and hearing voices that don't sound like voices. Work in our lab has shown that you can match heard voices to *point-light* videos of talking faces. For these purposes, we implemented the general methods used for our facial recognition experiments described in Chapter 8. Small, luminous points were placed on the face, lips, teeth, and tongue tip of our talkers. The talkers were filmed in the dark so the resultant video clips showed only white points moving against a black background. During the filming, the talkers uttered the sentence "The football game is over," a particularly easy sentence to lip-read. At a different time, our talkers were *audio* recorded uttering this sentence. Subjects were played the audio sentence of a talker and then saw two point-light talkers, one of which was the same as the talker they'd heard. Subjects were able to match the audio utterance to the same talker's point-light face at levels significantly above chance. Because the facial images only showed white points moving, it's unlikely that matches were based on talkers' age, size, or attractiveness. Instead, we think subjects might have based matches on something more specific: something that has also been suggested by another set of experiments.

In experiments conducted in another laboratory, point-light faces were tested for talker-based matches to audio stimuli that were highly reduced versions of speech recordings. These audio stimuli were composed of nothing more than three whistles that varied in pitch. The pitches of the whistles varied according to how critical parts of the recorded speech signals varied (formants and their transitions). Still, without knowing that these sounds were based on speech recordings, you'd say they sounded like nothing more than electronic whistles; similar maybe, to R2-D2's sounds. However, if you were told that the signals were based on speech, you'd start to hear them differently. Ultimately, you'd be able to make out syllables, words, and sentences in these whistle sounds. Research has also shown that you'd be able to recognize *whose* speech these signals were based on. And amazingly, you'd be able to listen to this whistle speech and match it to the correct point-light talker in a voice-face matching experiment. You could match voices to faces based on voices that didn't sound like voices and faces that didn't look like faces.

What do point-light talking faces and whistle voices have in common? As it turns out, many things. But most relevant to voice-face matching experiments is the fact that both point-light and whistle speech remove almost all types of the talker information usually considered useful for identifying people. Point-light speech removes the facial shape, skin tone, featural and configural information typically considered necessary for facial recognition. Whistle speech removes the fundamental voice pitch, timbre, and breathiness generally considered necessary for *voice* recognition.

But both point-light and whistle speech do retain one type of information that *can* identify a talker: idiosyncratic speaking *style* (articulatory style or idiolect). As you learned in Chapter 9, we all talk differently, even when speaking in the same dialect. (The one *possible* exception being identical twins at a young age.) When we speak, we have an idiosyncratic way of stringing together our consonants and vowels: we have a distinct style of articulating. It's thought that both point-light and whistle speech retain this speaking style information and *this* is what perceivers use to match point-light faces to whistle voices.

One interesting thing about speaking style information is that it is a dimension of speech that is not tied just to hearing, or just to seeing. It's a property of our articulation that is available both in heard *and* seen speech. This fact may help explain one of the most surprising findings we've ever observed in our lab—a finding described in the next section.

LIPREADING FROM A TALKER CAN HELP YOU HEAR THEM BETTER LATER

SUE THOMAS SAYS THAT she encounters people who are difficult to lip-read almost daily. Talkers may be difficult to lip-read because of their facial hair, or a more mumbling style, or in my case, too many smiles. But in most cases, additional experience with a talker allows her to better understand his or her speech. This is completely consistent with research conducted in both our lab and others, using hearing subjects. This means that like Thomas, you get better at lipreading the more familiar the talking face.

The same thing is true of auditory speech. Research shows that as you become more familiar with listening to a talker, you can better understand that person's speech when, for example, faced with a noisy environment.

It should be quickly mentioned that while both of these facts might seem intuitive (you're better at hearing and lipreading the speech of familiar people), these findings are relatively new and have actually forced major changes in speech perception theories.

What is it that subjects are *learning* when they get better at perceiving the speech of familiar talkers? It could very well be that they are becoming acquainted with a talker's *speaking style*. After all, it does seem like speaking style information is something we can use to recognize talkers, and to recognize their speech. But recall something interesting about speaking style information. It seems to be available in both heard and seen speech. This led our lab to make one of the wildest predictions we've ever made: experience with lipreading someone should help you *hear* that person better later.

Here's how we came up with the prediction. We knew that one way people get better at lipreading a familiar face is through gaining experience with his or her speaking style. The same thing is true of auditory speech: people get better at hearing a familiar person's speech *also* through experience with his or her speaking style. Now because speaking style is something that can be seen and heard, it shouldn't matter that much whether the experience is gained through lipreading or hearing. Thus, becoming familiar with a talker through lipreading should help a person *hear* the speech of that talker later.

And so we tried it, and it worked. We asked subjects who had no formal lipreading experience (but who had been screened for minimal lipreading ability) to lip-read sentences spoken from a talker for about an hour. The sentences were simple ("The flowers are on the table"), and subjects were generally able to correctly recognize three or four words in each sentence. After an hour of lipreading, subjects were asked to listen to the same types of sentences spoken against a background of noise. Unbeknownst to the subjects, the sentences were spoken by either the same talker whom they had just lip-read, or, for another group of subjects, a different talker. As they listened to the noisy sentences, subjects were asked to identify as many words as they could. We found that for those subjects who listened to the talker whom they had just lip-read, more words were understood than for those who heard a new talker. Their experience lipreading a talker transferred to help them *hear* that talker more easily.

We believe that this occurred because when they were lip-reading, subjects weren't just becoming familiar with how the talker *looked* as he spoke. They were, more generally, becoming familiar with *how he spoke*. This helped them perceive the talker's speech whether it was through sight or sound.

■　■　■　■　■

OVER 30 YEARS SINCE the McGurk effect was originally reported, it's become clear that speech perception is a multisensory function. It's likely that speech evolved to be seen as well as heard. This is supported by evidence that monkeys and chimpanzees are sensitive to audiovisual correspondences between heard calls and mouth configurations. And for integrating audiovisual calls, these primates use brain regions analogous to humans. There is also speculation that the world's languages have developed to take advantage of our auditory and visual speech skills. Languages typically show a complementary relationship between the audibility and visibility of their speech segments. Specifically, speech distinctions that are difficult to hear (*m* versus *n*) are easy to see and vice versa.

But research showing the importance and automaticity of multisensory speech has implications that go beyond our understanding of speech perception. As intimated, the research has helped motivate a reconsideration of the perceptual brain and how it handles multisensory input. This new conception is the topic of the final chapter.

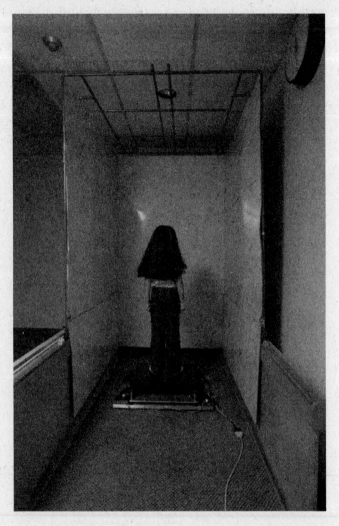

In the moving-room method, a small three-walled room is shifted forward and backward around a standing observer. The observer stands on a force plate to record her changes in posture. The method demonstrates how even our ability to stand still is strongly influenced by multiple senses.

CHAPTER 11

All of the Above

IN A WAY, THIS ENTIRE BOOK IS A LIE.

Well, at least the organization of this book is misleading. Like many books on perception, the chapters have been organized based on the functions associated with the sense organs: hearing, smelling, tasting, touching, seeing. But we now know that perception doesn't parse so easily this way. And you may have been getting this impression all along.

As you've learned, you can hear shapes and touch speech. You can taste odors and smell fear. You can see speech and hear space. You can touch flavor and smell symmetry. And you can taste scenes and hear faces. Many skills we've long assumed to be the purview of one sense are now known to also be performed by other senses.

Speech perception is one of the prototypic examples. You can perceive speech from seeing, touching, and listening to a face move. And while you're certainly best at hearing speech, there's an important way in which your brain doesn't care how it gets its speech input. And it might not even know. As the McGurk effect demonstrated, you are often unaware of the separate sensory contributions to what you perceive. When it comes to recognizing speech, your brain doesn't know if it's seeing or hearing.

But the more we're learning about the perceptual brain, the more we're realizing that this primacy of multisensory perception is true of many functions. This isn't to say that the brain doesn't make us aware that we're seeing versus hearing, and smelling versus touching. It's just that when it comes to the most important aspects of perception—guiding your action to effectively navigate the world—your brain often just doesn't know, or care, whether you're seeing, hearing, or touching the world. In fact, your brain could treat these types of input the same, because in some important respects, light and sound might *be* the same. But more on this later.

First, let's understand how widespread multisensory perception is. You've already learned about many examples. The rubber hand effects show that where you experience your hand being touched is strongly influenced by what you see. Both seeing and touching a face can influence the speech you hear. Many of the mirror system effects involve touch influencing your visual perception (holding a pen in your mouth affecting your perception of facial emotion). And of course, without multisensory influences, food wouldn't have much flavor at all.

But multisensory influences abound for even the simplest experiences. Depending on the types of images and sounds used, what you see and hear can influence each other in their perceived duration, intensity, direction of motion, and speed. In the aptly named *ventriloquist effect*, the perceived location of a sound is shown to depend on the location of a simultaneously seen light. This is actually an experience you encounter often when the action you see on a TV or movie screen captures the sound away from loudspeakers positioned some distance from the screen. Other research has shown that the number of simple flashes you see on a computer screen depends on how many simultaneous blips you hear. In another blip-based effect, whether you see two converging disks bounce off, or move through, each other depends on whether you hear a blip at the moment of contact (in which case you'd see a bounce).

Other simple cross-sensory effects involve touch. For example, being touched just as you see those two aforementioned converging disks make contact makes them seem to bounce off, rather than pass through, each other. And what you hear can influence what you feel. In a study reminiscent of the potato chip crunch experiment, the sounds of subjects' rubbing their hands together was picked up by microphone, modified, and played back to the subjects over headphones. By amplifying the treble (high-frequency components) of the rubbing sounds, most subjects reported experiencing that their hands *felt* drier. But touch can have its revenge.

Your sense of a sound's horizontal location can be influenced by whether you're simultaneously being touched on your right or left index finger.

And more complex, real-world sounds and sights can influence one another. The McGurk effect is a prototypic example of this in that the sight of complex speech movements can influence their sound. But other examples exist, many of them musical. In our lab, we found that the sight of someone plucking or bowing a cello can influence whether you hear a pluck or bow sound. Next, the sight of someone exaggerating the motions at the end of a marimba strike makes a marimba note sound longer. Also, the pitch difference heard between two sung notes is influenced by watching a singer sing two notes with either a greater or smaller pitch difference. And in another study conducted in our lab, the perceived loudness of a handclap is influenced by seeing handclaps of different apparent efforts.

Multisensory information for an event can also speed and enhance perception over unisensory information. In some of the most dramatic examples of this, two subliminal stimuli from different sensory modalities, when presented together, can both be noticed. Recall that this was observed in flavor perception when a subliminal sugar solution was sipped along with a subliminal cherry smell. Separately, they were undetectable, but together they tasted like cherry.

In other striking demonstrations, a simple stimulus presented to one sense can transfer to later influence another sense. A very recent example of this is noteworthy. First you need some background. Have you ever noticed that if you look steadily at a waterfall for a minute, and then look away to the nearby brush or rocks, these stationary objects seem to be moving *up*? In fact, this phenomenon is (creatively) called the *waterfall effect*. The effect has long been attributed to the fatiguing of the cells in your visual system that are sensitive to downward motion. But recent findings suggest otherwise. The waterfall effect can transfer from sight to touch. If you look steadily at a visual grating moving downward for 10 seconds, you'd then *feel* as if a stationary metal grating positioned on your finger were moving upward. It works the other way as well. If you first felt this small metal grating moving upward along your fingertip for 10 seconds, you'd then *see* a stationary visual grating move down. This means that the effects of perceiving movement through sight or touch are maintained in a form that is independent of these individual senses.

This finding fits an emerging story about how your senses work together. Simple perceptual effects we've always assumed were based on the physiology of a particular sense organ (your eyes, for example) are turning out to be shared and cross-influenced among multiple senses. One more example should drive this idea home.

Consider this simple reflex. Having your calf muscle subtly stimulated by a direct electrical pulse (through your skin) induces an unnoticeable, but measurable, electric response in that muscle. While too subtle to notice, the response seems reflexive in a way similar to how your lower leg lifts when the doctor taps your knee. And it would seem to be a simple, reflexive response constrained to your leg muscle: after all, a disembodied muscle would show the same response. But despite the seemingly reflexive and peripheral nature of this response, it can be changed by stimulation to your other senses. If, for example, you are presented a brief sound or a visible blip at the same time your muscle receives the electrical pulse, your muscular response is amplified. What would seem to be a simple, peripheral, and reflexive muscular reaction to a directly applied stimulus turns out to be a multimodal function. No sense, even the most peripheral, works on its own.

FOCUSING YOUR EYES CHANGES YOUR POSTURE

RON PITTMAN'S WORK HAS the potential to make you vomit. But he works very hard to ensure that doesn't happen. Pittman works for SimEx/Iwerks, the premier company for motion simulator amusement rides. Motion simulator rides involve an audience sitting in front of a large screen on which is projected a film shot from the point of view of a fast-moving car, roller coaster, spaceship, etc. Seeing a film in this way can give the audience a simulated but strong impression of their own motion. And the motion effect is enhanced by the sounds of movement, as well as the actual, albeit limited, movement of the seats themselves. Motion simulator rides are popular in theme parks, casinos, science museums, and zoos. Some of the best known are the *Star Tours* ride at Disneyland and (as of this writing) *The Simpsons Ride* at Universal Studios.

Ron Pittman's primary job at SimEx/Iwerks is to program the movements of the motion platform: the hydraulically controlled riser that supports anywhere from 10 to 50 seats. It's a job performed by maybe a dozen other people in the world, and Pittman takes it very seriously. "It's critical that I make the movement of the seats closely follow the movement the audience *sees*. If the felt and seen movements don't match, the audience can get uncomfortable and even feel a little sick," Pittman says. Fortunately, his 15 years of experience allow him to program a smooth yet exhilarating ride for his audience.

Motion platform programming is usually a hands-on procedure. Pittman typically flies out to the actual ride site after the seats have been installed and the film is ready for projection. To program the platform motion, he connects a joystick control to the platform system so he can control the movement of the platform as he watches the film. Depending on the ride, a seating platform can move along six different dimensions (up/down, forward/back, side-to-side, tilt forward/back, tilt side-to-side, turn left/right) and be displaced two feet in any direction. Obviously, this is nowhere near the amount of movement portrayed in the film (a flight through space; a roller coaster ride). So based on these motion constraints, Pittman must make the platform move in a way that tracks the motions simulated in the film in a compelling way. When done well, these movements enhance the audience's sense of motion. If, for example, the film simulates the perspective of a wide turn to the left, Pittman might shift the platform's front to face left, and then slowly shift the platform's back to swing—or fishtail—out to the right.

The platform motions Pittman creates with the joystick are digitally recorded so that they are synchronized with the film. Pittman can then play back these recorded motions to evaluate and fine-tune his motion sequence as he watches the film. Once satisfied with his work, the final motion sequence is saved and then played whenever the film is shown to an audience.

While a majority of the platform movement programming is done by joystick control, there are some motions that are just too subtle to produce this way. For these, he uses a computer editor that can more carefully adjust the platform's motions. "For example, we use a movement technique called 'washing back,'" Pittman explains. "These are seat movements that we don't want the audience to notice. They're often used to set up an extreme motion that the audience isn't expecting. If an upcoming action will involve the platform quickly going backwards, the platform will need to be imperceptibly moved forward so that it can be positioned to move back quickly. Now, it's very difficult to move the platform forward in an unnoticeable way using the joystick. So we go down to the computer editor to create that 'washing back' motion. It gives us much finer-grain control."

Of course, to give a compelling sense of flying through space or riding on a roller coaster, the film, itself, must be carefully designed. Pittman's colleague, Mike Frueh (vice president of SimEx/Iwerks) states that there are some rules of thumb that help the film create a good sense of visual motion. "We tend to find that the closer the viewpoint is to the ground, the better the sense of fast motion. The perspective of an airplane or helicopter pilot doesn't work quite as well as seeing the ground rush close by from a car or roller coaster. Having visual texture rush by near the viewpoint of the audience seems to give the best sense of speed and excitement." Frueh believes that another consideration is the consistency of the viewpoint. "It's important that the film shows a pretty consistent perspective for the motion experience to work. We've recently been working with clips from feature films that were not designed to be in motion rides. These films often contain fast edits that quickly change the perspective. The most recent one involves a car race. There are a lot of shots from the perspective of the driver, but every four seconds or so, that perspective changes. This turns out to be disorienting for the audience, especially if they're seated near the screen. And it weakens the motion effects."

But with a well-designed film along with Ron Pittman's seat-platform programming, a motion simulator ride can be an exhilarating experience. And it can be more comfortable for audience members feeling too old for *real* roller coasters. These simulator rides work based on a number of perceptual principles, one of which is the multisensory nature of perceived self-motion. This principle has been studied in perceptual laboratories for some time.

■   ■   ■   ■   ■

IMAGINE BEING IN THIS EXPERIMENT. You're told to walk into what looks like a small room, the size of a walk-in closet. The interior walls of the room are covered with fancy wallpaper. Your task is simply to stand still in the middle of the room and look forward. A thin, light wire is attached to your shoulders so that your posture can be monitored.

You take a deep breath, relax, and look forward. And you try to stand still. But very quickly, you realize that you're having trouble doing so. You feel yourself subtly swaying back and forth and have no idea why. And then your swaying stops. You hear the experimenter tell you that the next trial will begin soon, and you should again just stand as still as possible. You vow to yourself to stay still this time. But to no avail. Once the trial starts, you again feel yourself swaying back and forth, but this time not so subtly. Now you realize why this is happening. The entire room—all four walls and the ceiling—is slowly swinging back and forth around you. Only the floor is stationary. But knowing this doesn't help. As long as your eyes are opened, you're subtly swaying along with the room. What you see affects how you stand.

And what you see can affect what you feel you're doing. With enough movement of the room, not only would your body sway, but you'd also have a strong feeling that your position were changing. You'd feel like you were moving forward and back a few feet and may even feel compelled to look at your feet to ensure that you were still standing in one place.

Clearly, this experiment works on the same principle as Ron Pittman and Mike Frueh's motion simulator rides. However, the effects are much more subtle. In the moving-room experiments the movement of the visual scene can be very slight and still induce an experience of motion—and a recoordination of your body. Regardless, the principle is the same. Your experience of your position and how it changes can be strongly influenced by what you see. This is especially true if what you see is the type of visual information (global optic flow) that is produced when you move.

Whenever you move, there is a flow of *all* the visual texture around your position. If you move forward, for example, then the visual texture moves backward around you. Importantly, this type of universal visual flow occurs *only* when you're moving. It's a lawful, trustworthy type of information, which when your brain detects, informs you of, and *only* of, your own motion. That is, unless you're in a moving room or on a motion simulator ride. And that's what makes those rides effective. Your brain has evolved to use universal visual flow information to know you're moving. So when it detects this information in a simulator, you experience motion.

A substantial amount of research has explored the different parameters allowing these effects to work. It turns out that a relatively small visual image can induce an experience of self-motion. This allows you to feel some motion even when watching flight simulators and roller coaster simulators on your cell phone. Also, whether you see this flow information in your central or peripheral vision, you experience motion, and your posture is affected. As Mike Frueh and Ron Pittman might tell you, as long as visual flow is close up and consistent, you perceive yourself moving.

Even toddlers are influenced by visual flow. This is something I was able to observe myself in my colleague Mark Schmuckler's laboratory. One-year-old toddlers stood in a moving room whose floor was covered in cushions. And it's a good thing it was. As soon as the room moved back and forth one time, toddlers fell right over. Of course the cushions kept them from being injured in any way, and they usually got right back up and squealed for more.

But even the subtlest visual information can influence your posture. It turns out that just changing *where you focus* affects how you stand. In an experiment that did not involve a moving room, subjects were first asked simply to stand still and look at a target positioned 10 feet away. Then, without moving their eyes, they were asked to change their focus to a different target located just one foot away from their face. Measurements of subjects' postural sway during these conditions revealed that their degree of front-back and side-to-side movements were influenced by where they focused. In general, subjects swayed more when they were focusing on the far target.

This effect shows how strongly your vision integrates with your muscle-tendon sense, and balance sense, to control your posture. When you shift your visual focus in this way, you're doing little more than changing the shape of your eyes' lenses one-sixteenth of an inch. But just this much change is enough to induce a recoordination of your body. This shows just how pervasive multisensory integration can be.

Not only does what you see change your posture, but what you hear changes it too. If you were blindfolded and heard sounds shifting around you, your standing posture would change in predictable ways. This would occur whether the sounds were moving in the front-back direction, from side-to-side, or rotating around you. In addition, you'd often have a sense that you, yourself, were moving when listening to these sounds. There's also evidence that your own sense of motion would be substantially enhanced if the sounds were accompanied by touch stimulation synchronized with the sounds. If, for example, you were hearing sounds shift from front to back as you sat on a shaking chair, you'd have a more vivid sense of your own illusory forward motion than if the chair weren't shaking—a phenomenon that would be completely unsurprising to Ron Pittman. But the phenomenon also demonstrates the degree to which even something as basic as your sense of your own position is based on the combination of your senses.

In addition to informing your basic perceptual skills, the combination of your senses can inform perceptual skills that are, shall we say, exotic.

You Use Multiple Senses to Perceive Fertility and Attractiveness

If you happen to be a female stripper, here's some advice: it pays for you to work on the nights just before you ovulate.

Or so the research suggests. A group of strippers anonymously provided daily records of their tips per shift, including the number of lap dances provided to customers—where most money is made. These women, who were not on birth control, also provided daily information about their menstrual cycles. Over two-months' time, the women reported making 25 to 45 percent more money just before they ovulated—their *estrus* phase.

These types of results counter the long-held assumption that, unlike most other mammals, human females hide their estrus. There have been various explanations of this assumption including that hidden estrus might create the appearance of constant sexual receptivity, thereby facilitating long-term bonding and paternal care of offspring.

But, this stripper experiment, as well as a handful of other recent studies, suggests that estrus is *not* hidden and can in fact influence mating-related behavior. The other studies showed that men act more jealous and are more guarding of their (especially new) female companions during estrus.

So do strippers in estrus make more money because they somehow *advertise* estrus? If so, it seems clear that estrus is being communicated in an implicit way. Certainly men are not consciously aware of when estrus is occurring in females. And most women are typically unaware of the *exact* time of their estrus—a fact evidenced by the careful temperature monitoring often used by women trying to get pregnant. To a great degree, then, our awareness of estrus is an implicit perceptual skill.

But recent research is showing that estrus may, in fact, be implicitly communicated to *multiple* sensory systems simultaneously. As you learned in Chapter 4, men find scent samples of women in estrus to be more attractive and sexy than samples taken from the same women during non-estrus phases. And it seems that estrus can be implicitly perceived through visual means as well. It turns out that during her estrus, a woman's face is rated as a bit more attractive than when her face is seen at other points of her cycle. This may be a result of some features (ears, fingers, breasts) actually becoming slightly more *symmetric* during a woman's estrus. There's also evidence that a woman's skin color lightens at that time. Women during estrus also show a subtly decreased waist-to-hip ratio—a characteristic that could in principle be more attractive to potential mates. Men often find a low waist-to-hip ratio an attractive feature in women.

Estrus may also be visually communicated through how a woman chooses to dress at that time. Thirty college-aged women were photographed twice during the month, once during their estrus phase and once during non-estrus (as determined by urine tests). These photographs showed the women's entire bodies, in a neutral position with their hands at their sides. To remove any influence of facial attractiveness, however, the women's faces were obstructed by a black oval in the photograph. These women were photographed wearing the clothes they had chosen to wear for the day and were completely unaware of the purpose of the study. The photographs were shown to male and female judges also naïve to the study's purpose. The photographs were shown in pairs, so that each woman was shown in the two phases of her cycle. The judges were asked in which of the two photos the woman was trying to look more attractive. At levels significantly higher than chance, the judges chose the woman's photograph taken during the estrus phase. Further, the closer to actual ovulation the woman's estrus photo was taken, the more likely that photo

All of the Above   277

was chosen by the judges. A follow-up experiment found similar results with another group of college-aged women from a different campus.

In these studies, the women were unaware that they were dressing more provocatively during their fertile (and prefertile) periods. Like a woman's odor, estrus may be visually communicated at an implicit level.

Not only can estrus be implicitly communicated through scent and sight, but it can also be conveyed through sound. It seems that a woman's voice subtly changes during her estrus and these changes make her sound more attractive. In these studies, women were recorded during both estrus and non-estrus days, uttering simple sentences or just counting from 1 to 10. These recordings were then played to male and female judges who were asked to rate the voices on attractiveness. Judges generally found women's voices more attractive during estrus. Other research has examined the types of acoustical changes occurring in a woman's voice over her cycle. It turns out that during estrus, a woman's voice rises in pitch a small but detectable amount.

Further, there is evidence that women in estrus have higher verbal fluency and creativity. They also self-report more flirtatious behavior. These changes seem to occur without most women's explicit awareness of their estrus phase or, at the very least, that their estrus has in any way changed their behavior. (Subjects never guessed the true purpose of any of these studies.)

Most of the physiological changes occurring during estrus can be associated with increased femininity. Women look, smell, and sound more feminine during estrus. Regardless, to the degree that estrus can be implicitly conveyed to others, it seems to do so through multisensory means. This may be one way that estrus stays hidden from our conscious awareness. On their own, the individual unisensory cues are far from conspicuous. But perhaps together, presented simultaneously across our senses, the information for estrus may be sufficiently, albeit still implicitly, usable. Information for estrus could then subtly change our mating-related behavior—whether in the lab, real-world, or "gentlemen's club."

■ ■ ■ ■ ■

REPRODUCTIVE POTENTIAL MAY ALSO be recognizable through multiple senses. Recall that facial attractiveness has been thought to convey reproductive potential via its symmetry, youthfulness, and "average" characteristics. Research is now showing that pretty people not only look attractive, but they smell and sound attractive too. It's long been known that ratings of a woman's facial and body attractiveness are correlated. But these visible dimensions have recently been shown be related to vocal characteristics. Interestingly, many of the studies use voice samples that are very simple, including nothing more than vowel sounds. Still, audio attractiveness ratings of these simple utterances relate strongly to the rated visible attractiveness of the faces of women who produced those utterances. These same utterances also related closely to the women's rated body shape attractiveness. And other research has found a relationship between women's facial attractiveness and odor attractiveness. Put simply, pretty voices come from pretty faces connected to pretty bodies that smell pretty.

The relationship between facial attractiveness and vocal attractiveness has been examined using those composite faces you learned about in Chapter 8. Recall that composite faces allow for the digital averaging of a large set of faces. These faces can be of a group of individuals with a particular trait (extroversion, openness to sex without love). And the facial compositing removes idiosyncratic traits of any one face, but retains the overall average features of that group.

To determine how voice attractiveness is related to facial attractiveness, 100 undergraduate women were recorded uttering simple vowels (*ee, ah, oo*) and photographed. The recordings were then played to men asked to rate the attractiveness of each of the voices. It turned out that the *pitch* of the women's voices largely determined how attractive they were rated. Specifically, higher-pitched voices were considered more attractive than lower-pitched voices, a finding that has been replicated in a number of labs. Based on the ratings, the faces of the 25 women with the highest-pitched voices were made into a composite, as were the faces of the 25 women with the lowest-pitched voices. When men were shown these two composite faces, they rated the composite from the high-pitched voice group as more attractive than that from the low-pitched group. This means that, in a sense, you can perceive an attractive voice by sight.

So what is it that underlies this cross-sensory female attractiveness? There's evidence that the common characteristics are femininity and, possibly, youthfulness. Higher-pitched voices are considered not only more attractive, but also more feminine and youthful, as are attractive faces, bodies, and odors. Both femininity and youthfulness, in turn, are related to a woman's estrogen levels, which can have a direct bearing on her reproductive development and health. It could be, then, that a woman's reproductive potential is signaled in her attractiveness—which can be apprehended in a multisensory manner, by our supremely multisensory brains.

YOUR BRAIN IS DESIGNED AROUND MULTISENSORY INPUT

YOU'VE ALREADY LEARNED MUCH about the cross-sensory flexibility of the perceptual brain. With sensory loss, the visual brain areas can respond to sound and touch, and the auditory brain can respond to touch and sight. This flexibility can occur whether the sensory loss is early or late onset. And even short-term sensory loss—as in blindfolding and touch-based neuroplasticity research—can show cross-sensory recruitment of brain regions. But there is also evidence that very minor sensory decrements can create compensatory perceptual plasticity. For example, it turns out that if you are nearsighted and wear glasses, you are a bit better at echolocating than most sighted people. It's thought that because glasses leave the visual periphery uncorrected, the spatial hearing of bespectacled individuals may be subtly enhanced. This effect shows that compensatory neuroplasticity is ready to offset even the smallest sensory reduction.

In a particularly stunning example of cross-sensory brain plasticity in animals, researchers have shown that connecting the eyes to the auditory brain of an infant ferret can induce that area to respond to visual input. Conversely, connecting the ferret's ears to its visual brain can create auditory responsiveness in that area. Using behavior-conditioning techniques with the ferrets, both of these studies showed that whether these animals *experienced* stimulation as light or sound depended on whether their ears or eyes were stimulated—*not* which brain region was reactive.

What does this extreme plasticity suggest? For many researchers, it's a hint that the perceptual brain is designed *around* multisensory input.

This, in turn, may mean that there is constant cross-talk among the senses, even in the once-presumed sensory-specific brain regions.

Until about 10 years ago, theories of multisensory perception assumed that cross-sensory influences occurred in the brain upstream from the separate brain regions thought dedicated to the individual senses. When you're talking with your friends over lunch, the sound of their voices was thought to be registered in your auditory brain, and the sight of their face speaking was thought to be registered in your visual brain. Any integration and enhancement of their seen face on the speech you heard was thought to occur at higher brain regions (association cortex).

But this conception of the perceptual brain is changing. You've already learned how lipreading can activate early parts of your hearing brain. This suggests that multisensory cross-talk occurs almost as soon as the speech input hits the brain. But other examples are emerging which show early cross-sensory influences. For example, the presence of visible events such as paper ripping, cup tapping, and wood sawing can change activity in your auditory brain relative to when simply hearing these sounds.

With regard to touch and vision, you've already learned that when you feel for your keys in your pocket, or perform some other spatial touch task, you make use of *visual* brain areas. In fact, transient brain lesions in your visual brain can severely disrupt your spatial touch abilities. Related research has shown that *identifying* certain objects by touch also induces activity in your visual brain. Other experiments show that the *position* of a touched target, whether to the right or left of your body, induces activity in the area of your visual brain responsive to the same *visual* locations. This could mean that at least portions of your visual brain are receptive to object location whether detected through touch or sight.

The long-held concept of the perceptual brain being composed of separate sense regions is being overturned. Your brain seems designed around multisensory input, and much of it doesn't care through which sense information comes. Your brain wants to know about the world—not about light or sound, as such.

■   ■   ■   ■   ■

THE RECENT EVIDENCE FOR the ubiquity of multisensory influences in perception, and the multisensory design of the brain, have led some of us to start entertaining a radical idea about perception. It could be that in an important way, the relevant input to your senses is the same whether detected by your eyes or ears. For sure, your eyes detect light, and your ears detect sound. But the same, abstract *patterns of information* are often observed across these types of input.

Consider this example. In Chapter 2 you learned how beep baseball players are able to anticipate the path of a beeping ball or their contact with the buzzing base. There was something special in the changing nature of these sounds that allowed the players to anticipate when to dive for the ball and when to open their arms to contact the vertical base. As the sound sources approached the players, there was a systematic change in, for example, the loudness that allowed the players to anticipate the location of the ball and base. Recall that this change in loudness can roughly be characterized as exponential such that it increases relatively slowly at first and then increases more rapidly as the sound source gets near (similar information has been shown to exist for echolocating approaching surfaces).

It turns out that the same thing happens when *seeing* an approaching object. Imagine that you're playing catch with a beeping baseball, but with your eyes open. The ball is thrown to you. If you think about how the ball's image is projected onto the back of your eyes (roughly speaking), you'll realize that this projection increases in size as the ball approaches. Here is the amazing part: as the ball gets closer, the *rate of change* of the image size on your eye is the *same* as the rate of change in the loudness of the beep (given certain assumptions). The rate of change is the same whether it's in the increasing image size or in the increasing beep loudness. In this way, the relevant information is the same in light and sound.

This could be a good thing for your brain. If it is this pattern of change your brain has been designed to detect, then it shouldn't much matter to your brain whether the pattern is conveyed through sight or sound. Thus, for perceiving *time to arrival*, your brain might not care about your eyes and ears being separate sensory inputs.

We conducted an initial experiment to test this possibility. We video and audio recorded a car approaching and passing an observation point at different speeds. The recordings were edited so that different vanishing positions could be established, and so that the sensory modality of the presentation could be changed for the entirety or parts of the car's approach. Conditions were designed so that subjects would both hear and see the car's entire approach or perceive the approach through hearing, then seeing, then hearing, then seeing, etc. This last condition, in which the audio and visual signals alternated (every half second), was the most direct test of whether the relevant audio and visual information were treated the same. Alternating the audio and video in this way created a stimulus with *discontinuous* sensory input (*audio, then visual, then audio, etc.*) but *continuous* time-to-arrival information. We figured that if perceivers attend to the time-to-arrival information as such, then it shouldn't matter that the approaching car alternated between being seen and being heard: subjects should do as well as when the stimulus was made up of full, intact audiovisual signals.

And this is exactly what we found. Our results revealed that despite the odd alternation of audio and visual signals, observers were as accurate in their time-to-arrival judgments with this stimulus as with the audiovisually intact stimulus. (Additional control stimuli were tested to ensure that it is the continuous presence of time-to-arrival information that supports performance accuracy.)

The idea that the relevant perceptual information takes a similar form across light and sound has also been explored in the context of speech perception. Some examples are intuitive. Try articulating *aba* repetitively at a smooth, comfortable pace. Now try doing the same thing but with a slightly faster speed. The rate at which you produce *aba* will be reflected in the *rate* with which the loudness and timbre of the sound changes. And this exact rate will also be reflected in the visual information for your lip movements. Thus, a brain working to perceive speaking pace would do well to search for this same rate information regardless of whether it's available in sound or light.

And there is now evidence that cross-sensory informational correspondences for speech go well beyond this intuitive example. Measurements taken from the front of a talking face have shown that the visible movements of the lips, tongue, and face correspond closely to certain aspects of the heard speech signal. In fact, understandable auditory speech can be synthesized based on these measurements of visible speech. This could suggest that the same dimensions are informative across the senses. Other similar examples of cross-sensory information take a more abstract form. Recall that both audio and visual speech signals can be reduced to nothing but their dynamic, changing attributes. This was demonstrated with the whistle speech and point-light techniques described in Chapter 10. Despite being severely reduced speech signals, both types contained enough dynamic, changing structure to be informative about what is said and who is saying it. A perceptual brain sensitive to these dynamic aspects of speech signals may not need to distinguish whether this information is contained in sound or light.

You Have a Type of Synesthesia

Zana Devitto is red and her sister is green. At least they've always referred to each other this way. These colors have nothing to do with their hair color or clothes. Instead these colors are based on their names—the first letters of their first names to be exact. Devitto and her sister are *letter-color* synesthetes. They experience images of specific colors when they look at letters or numbers. Devitto describes the experience: "It's not as if the letters look printed in different colored inks. The colors are less distinct, and seem more like they're surrounding the letters, or being projected onto them. Either way, the colors are inseparable from their *specific* letters." For example, for Devitto, O's are always white, P's are green, and R's are purple. And if a letter changes, the color changes along with it. If Devitto were to change a P into an R by drawing a single slanted line, the letter's color would spontaneously change from green to purple.

Synesthesia is a systematic and involuntary sensory experience induced by an unrelated stimulus. For Zana Devitto, seeing a letter or number induces a color. But there are many varieties of synesthesia. For many synesthetes, *sounds* can induce color sensations. These sensations, while amorphous and ethereal, often seem to have a definite spatial location. Other synesthetes experience these color images upon hearing or reading the days of the week. Less common is the synesthesia in which hearing certain words induces distinct flavors in the mouth. And in what is likely the rarest form, tasting a flavor can produce tactile feelings of shape. Devitto's letter-color synesthesia is one of the most common. Also common is the fact that she had no idea her experiences are unique.

"When we were kids, my sister and I discussed the fact that we saw letters as having colors. But we thought that everyone did! It wasn't until
I was in college and learned about synesthesia that I realized we were
unique," she recalls. Also common to Devitto's type of synesthesia is the
fact that the same colors have been associated with the same letters for
as long as she can remember. Her *P*'s have always appeared green, her
O's white, and her *S*'s, reddish-purple. Her sister has many of the same
letter-color correspondences, but not all. While, Zana Devitto's *P*'s are
dark green, her sister's are light green.

The estimated number of individuals in the population suspected to
have some form of congenital synesthesia ranges from 1 in 20,000 to 1
in 23. Synesthesia is known to run in families, as evidenced by Devitto
and her sister as well as the fact that their grandmother and cousin are
synesthetes. Devitto and all of her synesthetic relatives also share a creative streak. Devitto herself is an avid stained-glass hobbyist and her
grandmother is a successful painter. Devitto believes that synesthesia
may inform her own art by allowing her to experience colors as having
"personalities." In fact, synesthetes in the general population are more
likely to have some artistic profession or avocation. And there are some
well-known artists and musicians who have either been confirmed or
suspected to have synesthesia. These include the artists David Hockney
and Wassily Kandinsky and musicians Franz Liszt, Duke Ellington, and
John Mayer.

Early explanations of synesthesia considered the experiences as little
more than fanciful introspection or robust metaphors based on remembered associations. However, the last 20 years have provided compelling
evidence for synesthesia's legitimacy. Synesthetes perform differently on
perceptual tests in ways that would be nearly impossible to fake. Further, brain imaging has shown differences in how synesthetes respond
to their inducing stimuli. When presented letters or numbers, the brains
of letter-color synesthetes show much greater activity in areas associated
with color perception.

The more recent theories assume some type of genetically based, neurophysiological explanation for synesthesia. One current theory is that
synesthetes have retained low-level, cross-sensory connections that for
most of us have been pruned away since infancy. Another theory is that
synesthetes have normal neural connections but lack some of the inhibit-

ing processes that usually eliminate certain signals from coursing across sensory brain areas. Regardless, these neurophysiological influences do more than integrate the senses—we all have those. These neurophysiological influences of synesthetes induce spontaneous sensory images based on some other stimulus.

But while a neurophysiological basis to synesthesia may be likely, the specific *associations* experienced by synesthetes may be based on their experience. For example, children are often first presented letters as colored, in the form of, say, refrigerator magnets or puzzle pieces. Very recent examinations of a group of synesthetes have found that their specific letter-color associations did turn out to be based on their childhood exposure to a particular set of letter magnets (which they still owned). Common childhood exposure to colored letters may explain why Devitto and her sister have most of the same letter-color associations. In fact, it turns out that *many* letter-color synesthetes share the same associations. Like Devitto and her sister, many synesthetes see *P*'s as green and *O*'s as white. Thus, the current wisdom is that while being a synesthete is related to some genetically based, neurophysiological anomaly, the specific associations experienced by a synesthete may very well be based on associations established in childhood.

But what about you? Probability suggests that you aren't a synesthete like Devitto and her sister. But according to one of the neurophysiological theories described above, you once were—before development pruned the extra connections. In fact, by some accounts, all infants are synesthetes. And according to the other neurophysiological theory, you have all of the hardware required to be a synesthete, your brain just inhibits some of the lowest-level cross-sensory influences. Either way, your brain is not too different from that of a synesthete. So do you have *any* type of synesthesia? New research suggests that you might.

For example, it turns out that many of the letter-color associations reported by Zana Devitto and other synesthetes are the same associations you'd make if asked. When a group of nonsynesthetes was asked to write down the first color that came to mind for the 26 letters of the alphabet, they often chose a majority of the same colors that letter-color synesthetes identified. Both groups often associated *A* with red, *B* with blue, C with yellow, and X with black. Of course, associations for both synesthetes and nonsynesthetes may be based, in part, on experience first learning letters in childhood, so this may not seem that surprising.

What may be more surprising is that your perceived correspondences between dimensions of colors *and sound* also follow those of synesthetes. If you were asked to choose colors that correspond to heard musical instrument (piano, violin, bagpipe) notes and chords, many of your selections would match those of synesthetes. Like synesthetes, you would be likely to choose lighter colors for higher notes and dark colors for low notes. You and the synesthetes would also choose more vivid colors for piano tones and more diluted colors for violin tones. You'd also both be likely to choose brighter colors for louder sounds. You'd also choose many of the same types of image *shapes* to correspond to different sounds. Like a synesthete, you'd be more apt to depict a low sound with a rounded figure, and a high sound with a more angular figure.

Of course, your specific selections of image-sound correspondences would not be as consistent as those of a synesthete from session to session. This is no doubt because a synesthete's choices would be based on *absolute* correspondences between sounds and sights: high C is *always* yellow. In contrast, your choices would be based on *relative* correspondences: you might assign the higher of two notes as yellow regardless of which specific notes you heard.

It may be this consistency in correspondences that facilitates many synesthetes' artistic and musical prowess. In fact, the artistic advantage of synesthetes was recently demonstrated in a colorful experiment. Five sound-color synesthetes (with little formal art or music training) were asked to describe their visual sensations and how those sensations changed as they listened to violin and cello sounds. From these descriptions, an artist created animations of the visual images. (These wonderful animations can be viewed by searching youtube.com for "animations samantha moore.") The animations were then paired with the original inducing sounds to create a set of audiovisual stimuli depicting the synesthetes' experiences. In addition, three sets of audiovisual stimuli which distorted this experience were created. In the distorted sets, the animations appeared upside down, with the wrong (complementary) color composition, or paired with the wrong (noninducing) sound. All of these audiovisual clips were shown to nonsynesthete subjects who were asked to rate how much they liked them. It turns out that the subjects liked the audiovisual clips that reflected the *actual* synesthetes' experiences more than the other clips. This suggests that while you may not have spontaneous visions when you hear sounds, you implicitly know what those visions should be like.

In a follow-up experiment, five *non*synesthetes were asked to draw figures that depicted the same cello and violin sounds the synesthetes heard. They were also asked to describe how these figures changed. The artist then created animations based on these images and descriptions. A set of naïve subjects then heard the inducing sounds while they simultaneously watched two animations derived from a synesthete's and nonsynesthete's description. When asked which of the animations they preferred, subjects chose the synesthete-derived animations substantially more often.

So true synesthetes may have the aesthetic advantage of a lifetime of vivid, consistent cross-sensory correspondences. But there is hope for the rest of us, if we're willing to be hypnotized.

IMAGINE BEING IN THIS spooky experiment. You're first screened to ensure that you are a person who can be easily hypnotized. You're then asked to sit in a darkened room and relax. A trained hypnotist enters and asks you to slowly go into a deep, deep sleep. Once under hypnosis, you're asked to look at a computer screen on which you are shown single digits, 1–6. Each digit is shown in a specific color (1, red; 2, yellow; 3, green; etc.) against a black background. Each time you see a digit, you hear the hypnotist say (in a spooky voice): "Look at that color; this is the color of the digit X, and whenever you see, think, or imagine that digit, you will always perceive it in that color." You're being hypnotized into becoming a synesthete.

In fact, after the session, you would experience something very much like number-color synesthesia. You would report seeing the numbers 1–6 as being the colors you've been told to see. And because, while under hypnosis, you were told to forget about the hypnosis session, you would probably have little idea why you were seeing colored numbers. But more compelling evidence for your induced synesthesia comes from a laboratory demonstration. You're asked to sit at a computer monitor and watch for numbers. On each trial, you see a "+" target that then changes to a colored screen which either contains or does not contain a number presented in black typeface. You're asked to press one of two keys to indicate whether you see a number. It's an easy task . . . for nonsynesthetes. But now that you're a synesthete you'd have trouble with the task whenever a color-associated number (1) appears against a similarly colored background (red), despite the fact the number is presented in black typeface. You'd make more errors and respond more slowly on those trials.

The fact that some individuals can be hypnotized into being synesthetes (for a short period of time) has some interesting implications. First, it suggests that synesthesia can occur in individuals who have undergone the normal neural pruning that occurs after infancy. This could mean that the other neurophysiological explanation—that synesthesia is based on a lack of inhibition between sensory brain areas—is more viable. In fact, hypnosis is known to remove inhibitory influences in a number of other contexts. More generally, however, these hypnosis results add to the findings that we may all have some synesthetic potential and characteristics.

YOU ACTUALLY MAY EXPERIENCE one type of synesthesia all the time. You have a vivid, spontaneous experience of flavor when you smell the food in your mouth. Recall from Chapter 5 that the aromas that leave your mouth and enter the back of your olfactory system are always experienced as flavor. By some accounts, this is a cross-sensory experience that fully qualifies as synesthesia. Your experience of aroma-induced flavor is automatic, consistent (across your lifetime), and can even arise without there being any stimulus on your tongue.

The synesthetic tendencies in all of us have led some theorists to propose that there are actually two types of synesthesia. *Strong* synesthesia is the type experienced by Zana Devitto, in which vivid sensory experiences arise from an unrelated class of stimuli. *Weak* synesthesia, on the other hand, is everyone's sense of cross-sensory correspondences. As you've just learned, there is research showing that you have the same sense of these correspondences as a synesthete (high pitch: light shade). But this sense appears in more contexts than just laboratory experiments. It appears in your use and understanding of cross-sensory *metaphor*. We all understand what is meant by a "warm color" or a "loud tie." The meaning of these metaphors depends on a universal awareness of how temperature and sound correspond to visible color and pattern. The existence of these types of metaphors may reflect an implicit—or explicit—awareness of cross-sensory correspondences which qualify as a weak type of synesthesia.

Interestingly, there are aspects of metaphor that are similar to characteristics of *strong* synesthesia. For example, in most languages, there are many more metaphors that take the form of a sound qualifying a sight ("loud tie"), than a sight qualifying a sound ("bright thunder"). In an analogous way there are many more synesthetes for whom sound induces a color than color induces a sound. This could mean that common processes are at work in guiding synesthetic correspondences in both their strong and weak forms.

It is likely your sense of cross-sensory metaphor that allows Bob Mitchell to convey visible action in his organ accompaniments and for the first two notes of "Somewhere over the Rainbow" to implicitly represent a rainbow.

It is also a sense of cross-sensory metaphor that allows Eileen Kenny to match an aroma to a visible décor. Recall that in choosing an aroma to match a commercial site, she often selects odor components that appeal to the type of clientele frequenting the site (using *lavandin* in trendier hotel lobbies). But she also chooses aromatic components that match the *look* of a site. "Sometimes my choices will be pretty obvious. If a lobby has a lot of yellow tones, I might choose a lemon scent," she states. "But other times, it will be more subtle and metaphorical. Imagine a hotel lobby that has a large, frosted glass wall with a waterfall image etched into it. The lobby might also have a large polished chrome center table and polished marble floors. But behind the front desk, there is a painting which contains a very bright splash of red.

"For this lobby, I might choose a splash of mandarin or something a bit spicy to match the redness in the painting. To that, I might add something clean and cool to match the polished chrome and marble—a white floral note for example, maybe with a hint of white tea. And to match the frosted glass wall with the waterfall etching, I could add something fresh like an ocean spray aroma along with a touch of lavender." For other sites, Kenny designs a scent to *offset* a space's limitations. "When you walk through a long narrow hallway, you can start to feel claustrophobic. For that space, I might choose a more airy fragrance to give the space a more open feeling. We like to think of this as 'aroma as architecture.'"

While 20 years of scent designing have allowed Kenny to become especially proficient at matching scents to spaces, her success is dependent on your own cross-sensory awareness. Your sensitivity to cross-modal correspondences—your weak synesthesia—has likely allowed you to use a wide variety of aesthetic metaphors. But it may have provided much more.

YOUR SYNESTHESIA MAY HELP YOU WITH LANGUAGE

ACCORDING TO V.S. RAMACHANDRAN and his colleagues (and discussed in his fascinating book, *A Brief Tour of Human Consciousness*), our synesthetic tendencies may have been integral to the evolution of language. To understand how, imagine yourself in this alien experiment. You're shown two figures. One has a rounded, amoeboid shape—akin to the profile of cloud. The other is much more angular, with a profile composed of straight edges forming points. You're told that these figures are the symbols for the first two letters of the Martian alphabet. You then hear two names—"booba" and "kiki"—and are asked to guess which of these names refers to which shape. So which is the "booba" shape? If you're like 95 percent of people asked, you would say "booba" referred to the rounded, globular shape, and "kiki" to the more angular, pointed shape. In fact, you would have made these same guesses when you were two and a half years old.

Why do people overwhelmingly label the rounded figure "booba" and angular figure "kiki"? It could very well be that to your (weakly) synesthetic brain, these figures look like they sound. And they may look like how *you speak*. Consider how you articulate "booba." Your lips move in a relatively smooth, continuous way, and actually form a rounded shape in articulation of the *oo* vowel. Articulating "kiki" in contrast, involves more abrupt, sharp movements of your tongue back, emulating the more angular form of the "kiki" shape. Similar correspondences have been found for other nonsense words ("Baamoo/Kuhtay"; "Gogaa/Teetay"; "Mabooma/

Tuhkeetee") and rounded versus angular shapes. You perceive the correspondences between these words and shapes because of your implicit sense of how these words are articulated and the types of sound characteristics that result.

What does this mean for language? According to Ramachandran, it could mean that at a preliminary stage in the evolution of language (protolanguage), our weak synesthesia facilitated the ways sounds were mapped to objects. In other words, *words*. It could be that the emergence of early words were based, in part, on the cross-sensory correspondences that we all implicitly know. So rounded objects may have initially been assigned rounded sounds and angular objects may have been assigned sharper, more abrupt sounds. Further, this process may have been facilitated by none other than the *mirror systems* you learned about in Chapter 9. It could be that our natural inclination to imitate the things we perceive induced our hominid ancestors to articulate in ways that reflected what they saw. Our ancestors saw a rounded object and articulated in a rounded way: "boulder." They saw a straight, angular object and articulated in a sharper, more abrupt manner: "stick."

Unsurprisingly, this is a controversial idea. Most words of modern languages don't sound like the things they name, at least not in obvious ways. In fact, the supposed arbitrary relation between word sounds and meanings is thought to provide the unique *generative* power of human language. But, there are some fascinating exceptions that may show remnants of the synesthetic basis of words. The most obvious is onomatopoeia: instances where words sound like the sounds, or sound-creating actions, they describe (*woof, boom, meow*). It also appears that many words describing size show a cross-sensory correspondence similar to the "booba/kiki" example. In many languages, words referring to little things often contain sounds made by shrinking (narrowing) a part of the vocal tract: *petite, teeny, diminutive, little*. In contrast, words describing large things—*huge, enormous, humongous, large*—are produced with a more open vocal tract. Going a step further Ramachandran suggests that the sounds of some words could be by-products of articulatory movements subtly mirroring hand gestures. In many languages, the words we use to denote "you" are formed by protruding the lips outward (*tu, vous, thoo*) as if pointing away from oneself. The words used to refer to oneself (*me, moi, naan*), in contrast, often involve the lips moving inward toward oneself.

Speculative as this is, the notion that there's something other than a purely arbitrary relation between word sound and meaning does have other support. Some support comes from research showing that perceivers can pick up on the sound-meaning correspondences in unfamiliar languages. In one study, non-Japanese-speaking subjects were presented pairs of Japanese antonyms (*ui* and *shita*) and asked which of the words corresponded to the translated English antonym pair (*up* and *down*). Subjects could make these matches at better than chance levels despite the vast difference between English and Japanese words and their ostensive etymologies.

And very recent research suggests that your sensitivity to subtle sound-meaning correspondences facilitates your memory for words. Non-Japanese-speaking subjects listened to Japanese words and were told to remember their English translation: "*suppai* means 'sour.'" Unbeknownst to these subjects, half of the translations were false, so that the Japanese word was randomly assigned its English translation: "*ui* means 'sweet.'" After learning the words, a memory test was conducted in which subjects heard a spoken Japanese word and chose one of two printed English words as the correct translation: "Does *suppai* mean 'sour' or 'up'?"

While most subjects performed well on the memory test, the *speed* with which they responded depended on whether the translation they learned was true or false. Subjects were faster at choosing the correct word when its translation was a *true* translation ("sour" for *suppai*), than when its translation was false ("sweet" for *ui*). Of course, subjects never explicitly knew that some of the translations were true and some, false. But *implicitly*, subjects seemed to know which translations were more likely to be true and responded to them more quickly. And it could very well be that this implicit knowledge of true translations was based on subjects' sensitivity to correspondences between the Japanese words' sounds and meanings. These results could mean that Japanese, at least, contains words that have subtle correspondences between sound and meaning. And the results could suggest that you are implicitly sensitive to these types of correspondences and can even use them in your language learning.

Regardless of the speculative and controversial notion that languages possess a type of "sound symbolism," it could very well be that some words contain remnants of the synesthetic basis of early language. In fact, your own synesthetic tendencies may allow you to implicitly detect these correspondences. While you may not have spontaneous images like Zana Devitto experiences, your ingrained sense of cross-sensory correspondences could underlie your understanding of metaphor as it appears in your aesthetic and linguistic world.

WHETHER PERFORMING A SIMPLE skill like standing and looking, or something more exotic like detecting estrus, your perceptual experiences are fundamentally multisensory. This is a consequence of your brain, and perhaps the most relevant aspects of the input information itself. For a few of us, the supreme multisensory nature of perception has the interesting by-product of ephemeral visions. But for all of us, perception's multisensory nature helps provide a robust experience of the world, as well as a way to convey that experience in a vivid, poetic manner.

Epilogue

JUST BELOW YOUR AWARENESS, YOU LIVE IN A PARALLEL WORLD: a world where your evolutionary heritage (loosely speaking) keeps you safe. Your inner bat listens to the spaces you occupy. Your inner rabbit listens for threats and anticipates their approach. Your inner dog allows you to determine the location of smells and your inner mouse helps you implicitly use those smells to perceive family, fertility, and reproductive potential. Your inner spider allows you to feel things without directly touching them and your inner firefly helps you sync with people. Your inner monkey helps you recognize intent from faces and effectively mimic the behaviors of others. And your inner ferret allows all of these skills to be fine-tuned through the neuroplasticity inherent to nervous systems.

Of course, what makes this inner menagerie most effective is its ability to collaborate through the multisensory perception intrinsic to everything you do. But one can also imagine your inner menagerie, just below your awareness, looking up at you and your more conscious endeavors. It would no doubt be in awe of the fact that you can read about, and reflect upon, yourself. You are self-aware like no other animal. But the luxury of this self-awareness is made possible by that same inner menagerie and the implicit perceptual skills it engenders.

Perhaps some of your implicit skills emerge in your consciousness as *intuition*. You may feel it's intuition that leads you to like some people more than others. But this intuition may be subtly influenced by your implicit detection of their odor or how they inadvertently mimic your behavior. You may feel that intuition plays a role in your choice of an office or apartment. But that intuition may be informed by your implicit detection of how these spaces reflect sound. In fact, by some accounts, a vast majority of our behaviors are guided by unconscious rather than conscious skills. This is a controversial perspective. But one thing now seems undeniable: much of what we do is richly informed by implicit perception. As any child of the sixties will tell you, consciousness is over-rated.

But as it turns out, many of your implicit skills are within your conscious reach. You can learn to hear more details of reflected sound and taste the separate sensory components of flavor. You can refine your ability to match odor to décor or to feel the characteristics of a fish on your line. You can boost your awareness of how faces match voices or how someone's idiosyncratic characteristics can be mimicked for comic effect. You could even learn to perceive speech through touch if you so choose. It's as if your inner menagerie—with all its implicit skills—sits waiting to help you consciously harness your perceptual potential.

Notes

CHAPTER 1. *The Sounds of Silence*

6 The eighteenth-century philosopher Denis Diderot: Diderot, D. (1972). Letter on the blind for the use of those who see. In M. Jourain (Ed. and Trans.), *Diderot's early philosophical works* (pp. 68–141). New York: Burt Franklin (original work published 1749).

7 Other reports of the blind's ability to sense objects appeared: Hayes, S. P. (1935). *Facial vision or the sense of obstacles*. Watertown, MD: Perkins Institution.

7 The most prevalent theory held that the blind: Hayes (1935), *Facial vision or the sense of obstacles*.

7 A definitive test of facial vision was conducted: Supa, M., Cotzin, M., & Dallenbach, K. M. (1944). Facial vision: The perception of obstacles by the blind. *American Journal of Psychology, 57*(2), 133–183.

7 After Dallenback's team established that sound: Kellogg, W. (1962). Sonar system of the blind. *Science, 137,* 399–405; Rice, C. E. (1967). Human echo perception. *Science, 155,* 656–664.

7 Research on bats at the time: Griffin, D. R. (1958). *Listening in the dark*. New Haven: Yale University Press.

8 These properties include: Kellogg (1962), Sonar system of the blind; Rice (1967), Human echo perception.

8 In my own lab's research: Rosenblum, L. D., Gordon, M. S., & Jarquin, L. (2000). Echolocation by moving and stationary listeners. *Ecological Psychology, 12*(3), 181–206.

9 Like bats, you probably use the time delay: Schenkman, B. N. (1985). Human echolocation: A review of the literature and a theoretical analysis. *Uppsala Psychological Reports, 379.* 1–34.

9 However, these types of sound cues: Schenkman (1985), Human echolocation.

9 One of the most important is the sound wave: Schenkman (1985), Human echolocation.

10 Research shows that the perceptual brain: Pascual-Leone, A., Amedi, A., Fregni, F., & Merabet, L. B. (2005). The plastic human brain cortex. *Annual Reviews of Neuroscience, 28,* 377–401.

10 It has long been known that people who: Amedi, A. Merabet, L. B., Bermpohl, F., & Pascual-Leone, A. (2005). The occipital cortex in the blind: Lessons about plasticity and vision. *Current Directions in Psychological Science, 14*(6), 306–311.

11 But even individuals with adult-onset blindness: Burton, H. (2003). Visual cortex activity in early and late blind people. *Journal of Neuroscience, 23*, 4005–4011.

11 As you'll learn in Chapter 6, tapping a subject's finger: Pleger, B., Dinse, H. R., Ragert, P., Schwenkreis, P., Malin, J. P., Tegenthoff, M. (2001). Shifts in cortical representations predict human discrimination improvement. *Proceedings of the National Academy of Sciences USA, 98*, 12255–12260.

12 This type of echolocation, known as *passive echolocation*: e.g., Xitco, M. J., & Roitblat, Herbert L. (1996). Object recognition through eavesdropping: Passive echolocation in bottlenose dolphins. *Animal Learning & Behavior, 24*, 355–365.

14 Reflected sound helps your brain recognize the type of space: Robart, R. L., & Rosenblum, L. D. (2005). Hearing space: Identifying rooms by reflected sound. In H. Heft & K. L. Marsh (Eds.), *Studies in perception and action XIII* (pp. 152–156). Hillsdale, NJ: Lawrence Erlbaum Associates, Inc.

21 Studies have shown that we automatically incorporate: Hartmann, W. M., & Rakerd, B. (1999). Localization of sound in reverberant spaces, *Journal of the Acoustical Society of America, 105*, 1149.

21 When listeners are asked to judge the degree: Bradley, J. S., & Soulodre, G. A. (1995). Objective measures of listener envelopment. *Journal of the Acoustical Society of America, 98*, 2590–2597.

22 Your ability to hear space from reflected sound: Robart & Rosenblum (2005), Hearing space.

23 There is some scientific research consistent with this conclusion: Ashmead, D. H., Wall, R. S., Eaton, S. B, Ebinger, K. A., Snook-Hill, M. M., Guth, D. A., Yang, X. (1998). Echolocation reconsidered: Using spatial variations in the ambient sound field to guide locomotion. *Journal of Visual Impairment & Blindness, 92* (9), 615–632.

26 Recent research in my own lab shows that unpracticed: Gordon, M. S., & Rosenblum, L. D. (2004). Perception of acoustic sound-obstructing surfaces using body-scaled judgments. *Ecological Psychology, 16*, 87–113.

26 Our research shows that your brain can detect even more: Rosenblum, L. D., & Robart, R. L. (2007). Hearing silent shapes: Identifying the shape of a sound-obstructing surface. *Ecological Psychology, 19*, 351–366.

CHAPTER 2. *Perfect Pitches, Beeping Pitches*

30 Beep baseball has been played: Edward Bradley, personal communication, June 4, 2008.

32 There are a number of acoustic dimensions: Zahorik, P., Brungart, D. S., & Brokhorst, A. W. (2005). Auditory distance perception in humans: A summary of past and present research. *Acta Acoustica, 91*, 409–420.

32 Take the loudness increase as an example: Shaw, B. K., McGowan, R. S., & Turvey, M. T. (1991). An acoustic variable specifying time-to-contact. *Ecological Psychology, 3*(3), 253–261.

33 In fact, this sort of *time-to-arrival* information: Lee, D. N., van der Weel, F. R., Hitchcock, T., Matejowsky, E., and Pettigrew, J. D. (1992). Common principle of guidance by echolocation and vision. *Journal of Comparative Physiology, 171*, 563–571.

34 When you hear an approaching vs. receding or stationary sound source: Seifritz, E., Neuhoff, J. G., Bilecen, D., Scheffler, D., Mustovic, H., Schächinger, H., Elefante, R., & Di Salle, F. (2002). Neural processing of auditory 'looming' in the human brain. *Current Biology, 12*, 2147–2151.

34 The adaptive importance of this system: Maier, J. X., & Ghazanfar, A. A. (2007). Looming biases in monkey auditory cortex. *Journal of Neuroscience, 27*, 4093–4100.

34 Unsurprisingly, these monkeys also show: Ghazanfar, A. A., Neuhoff, J. G., & Logothetis, N. K. (2002). Auditory looming perception in rhesus monkeys. *Proceedings of the National Academy of Sciences USA, 99*(24), 15755–15757.

34 Four-month-old infants: Freiberg, K., Tually, K., & Crassini, B. (2001). Use of an auditory looming task to test infants' sensitivity to sound pressure level as an auditory distance cue. *British Journal of Developmental Psychology 19*, 1–10.

34 Research shows that most listeners: Neuhoff, J. G. (2004). Auditory motion and localization. In J. G. Neuhoff (Ed.), *Ecological Psychoacoustics* (pp. 87–111). New York: Academic Press.

35 The auditory approach research also shows: Rosenblum, L. D., Gordon, M. S., & Wuestefeld, A. P. (2000). Effects of performance feedback and feedback withdrawal on auditory looming perception. *Ecological Psychology, 12*(4), 273–291.

35 It is not surprising, then, that visually impaired listeners: Schiff, W., & Oldak, R. (1990). Accuracy of judging time to arrival: Effects of modality, trajectory, and gender. *Journal of Experimental Psychology: Human Perception and Performance, 16*(2), 303–316.

35 When you hear a sound *increasing* its intensity: Neuhoff, J. G. (1998). Perceptual bias for rising tones. *Nature, 395*(6698), 123–124.

35 In a related illusion, your perception of a sound: Neuhoff, J. G. (2001). An adaptive bias in the perception of looming auditory motion. *Ecological Psychology, 13*(2), 87–110.

36 the widespread negative effects of noise pollution: Coghlan, A. (2007). Dying for some quiet: The truth about noise pollution. *New Scientist*, August, 22.

36 We consistently find that listeners need approaching hybrid cars: Robart, R. L., & Rosenblum, L. D. (2009). Are hybrid cars too quiet? *Journal of the Acoustical Society of America, 125*(4), 2744.

39 But even if you've never met your thief: Li, X., Logan, R. J., & Pastore, R. E. (1991). Perception of acoustic source characteristics: Walking sounds. *Journal of the Acoustical Society of America, 90*, 3036–3049.

39 Related research shows that from listening to the footsteps: Pastore, R. E., Flint, J. D., Gaston, J. R., & Solomon, M. J. (2008). Auditory event perception: The source-perception loop for posture in human gait. *Perception & Psychophysics, 70*(1), 13–29.

39 Continuing your surveillance, if you happen to hear a familiar thief: Repp, B. H. (1987). The sound of two hands clapping: An exploratory study. *Journal of the Acoustical Society of America, 81*, 1100–1110.

40 The research suggests you'd easily recognize: Marcell, M., Malantanos, M., Leahy, C., & Comeaux, C. (2007). Identifying, rating, and remembering environmental sound events. *Behavior Research Methods, 39*(3), 561–569.

40 You also have laudable skills at auditorily identifying: Marcell, M., Barella, D., Greene, M., Kerr, E., & Rogers, S. (2000). Confrontation naming of environmental sounds. *Journal of Clinical and Experimental Neuropsychology, 22*(6), 830–864.

40 Impressively, your accuracy would not be substantially reduced: Ballas, J. A. (1993). Common factors in the identification of an assortment of brief everyday sounds. *Journal of Experimental Psychology: Human Perception and Performance, 19*(2), 250–267.

40 Research shows that listeners can determine the length: Carello, C., Anderson, K. L., & Peck, A. (1998). Perception of object length by sound. *Psychological Science, 9,* 211–214.

41 In one experiment, blindfolded listeners: Kunkler-Peck, A., & Turvey, M. T. (2000). Hearing shape. *Journal of Experimental Psychology: Human Perception and Performance, 1,* 279–294.

41 Research suggests that you are excellent at hearing the temporal details: Cabe, P., & Pittenger, J. B. (2000). Human sensitivity to acoustic information from vessel filling. *Journal of Experimental Psychology: Human Perception and Performance, 26,* 313–324.

43 But there are techniques that increase the chance of success: Torres, R. (2004). The art of in-tune singing. *Vocal Education Series.* Available online: http://www.mastersof harmony.org/vocal_education_series.

43 Tonal or timbral matching across members: Torres (2004), The art of in-tune singing.

43 A final trick used by the Masters of Harmony: Torres (2004), The art of in-tune singing.

45 In a recent study, individuals were randomly approached: Bella, S. D., Giguere, J. F., & Peretz, I. (2007). Singing proficiency in the general population. *Journal of the Acoustical Society of America, 121*(2), 1182–1189.

45 It turns out that inexperienced singers are notably worse: Amir, O., Amir, N., & Kishon-Rabin, L. (2003). The effect of superior auditory skills on vocal accuracy. *Journal of the Acoustical Society of America, 113*(2), 1102–1108.

45 Inexperienced singers don't typically have: Brown, W. S., Jr., Rothmann, H. B., & Sapienza, C. (2000). Perceptual and acoustic study of professionally trained versus untrained voices. *Journal of Voice, 14*(3), 301–309.

46 Roughly 1 in 10,000 adults: Deutch, D., Henthorn, T., & Dolson, M. (2004). Absolute pitch, speech, and tone language: Some experiments and a proposed framework. *Music Perception, 21*(3), 339–356.

46 A number of well-known musicians have either been: http://www.perfectpitch people.com, retrieved February 28, 2008.

46 As discovered by Daniel Levitin: Levitin, D. J. (1994). Absolute memory for musical pitch: Evidence for the production of learned melodies. *Perception and Psychophysics, 56,* 414–423.

46 When mothers are asked to sing their favorite: Bergeson, T., & Trehub, S. (2002). Absolute pitch and tempo in mothers' songs to infants. *Psychological Science, 13,* 17–26.

47 And even if you were asked to make a more explicit: Schellenberg, E. G., & Trehub, S. E. (2003). Good pitch memory is widespread. *Psychological Science, 14,* 262–266.

47 Even if you've never taken a lesson: Smith, J. D., Nelson, D. G., Grohskopf, L. A., & Appleton, T. (1994). What child is this? What interval was that? Familiar tunes and music perception in novice listeners. *Cognition, 52*(1), 23–54.

49 These qualities include how melodic themes: Bigand, E., & Poulin-Charronnat, B. (2006). Are we "experienced listeners"? A review of the musical capacities that do not depend on formal musical training. *Cognition, 100,* 100–130.

49 You may not explicitly know your romantic from your neoromantic: Bella, S. D., & Peretz, I. (2005). Differentiation of classical music requires little learning but rhythm. *Cognition, 96,* B65–B78.

49 But before you flaunt your newfound skill: Chase, A. (2001). Music discrimination by carp (Cyprinus carpio). *Animal Learning and Behavior, 29*(4), 336–353.

49 but sparrows can discriminate: Watanabe, S., & Sato, K. (1999). Discriminative stimulus properties of music in Java sparrows. *Behavioural Processes, 47,* 53–57.

49 The fact that musical styles: Bella & Peretz (2005), Differentiation of classical music.

50 But amazingly, you are implicitly sensitive to this rule system. Bigand, E., D'Adamo, D. A., Poulin-Charronnat, B. (in revision). The implicit learning of twelve-tone music.

52 Research suggests that you heard music: James, D. K., Spencer, C. J., & Stepsis, B. W. (2002). Fetal learning: A prospective randomized controlled study. *Ultrasound in Obstetrics and Gynecology, 20*, 431–438.

52 By the age of four months: Zentner, M. R., & Kagan, J. (1998). Infants' perception of consonance and dissonance in music. *Infant Behavior and Development, 21,* 483–492.

53 Research suggests that by the age of four years: Gregory, A. H., Worrall, L., & Sarge, A. (1996). The development of emotional responses to music in young children. *Motivation and Emotion, 20,* 341–348.

53 By six, you likely perfected: Cunningham, J. G., & Sterling, R. S. (1988). Developmental change in the understanding of affective meaning in music. *Motivation and Emotion, 12,* 399–413.

53 Research suggests that you can successfully: Peretz, I., Gagnon, L., & Bouchard, B. (1998). Music and emotion: Perceptual determinants, immediacy, and isolation after brain damage. *Cognition, 68*(2), 111–141.

53 Brain scans of listeners hearing major versus minor: Pallesen, K. J., Brattico, E., Bailey, C., Korvenjoha, A., Koivisto, A. G., & Synnove, C. (2005). Emotion processing of major, minor, and dissonant chords: A functional magnetic resonance imaging study. *Annals New York Academy of Sciences, 1060,* 450–453.

53 While these children have a very difficult time: Heaton, P., Hermelin, B., & Pring, L. (1999). Can children with autistic spectrum disorders perceive affect in music? An experimental investigation. *Psychological Medicine, 29,* 1405–1410.

53 Listening to music produces activity in brain areas: Peretz, I. (2006). The nature of music from a biological perspective. *Cognition, 100,* 1–32.

53 The emotional responses can appear as measurable: Krumhansl, C. L. (1997). An exploratory study of musical emotions and psychophysiology. *Canadian Journal of Experimental Psychology, 51,* 336–352.

54 Finally, you may not be surprised to learn: Peretz (2006), The nature of music.

54 In fact, your ability to interpret musical emotion: Juslin, P. N., & Laukka, P. (2003). Communication of emotions in vocal expression and music performance: Different channels, same code? *Psychological Bulletin, 129*(5), 770–814.

55 When watching an animated film portraying: Heider, F., & Simmel, M. (1944). An experimental study of apparent behavior. *American Journal of Psychology, 57,* 243–259.

55 Similar findings have been reported for more elaborate: Bolivar, V., Cohen, A. J., & Fentress, J. (1994). Semantic and formal congruency in music and motion pictures: Effects on the interpretation of visual action. *Psychomusicology, 13,* 28–59.

55 This research suggests that regardless: Lipscomb, S. D., & Kendall, R. (1994). Perceptual judgment of the relationship between musical and visual components in film, *Psychomusicology 13,* 60–98.

CHAPTER 3. *You Smell like a Dog*

60 This human scent-tracking experiment: Porter, J., Craven, B., Khan, R. M., Chang S. J., Kang, I., Judkewicz, B., Volpe, J., Settles, G., & Sobel, N. (2007). Mechanisms of scent-tracking in humans. *Nature Neuroscience, 10,* 27–29.

62 There are many reasons for a dog's advantage: Shepherd, G. M. (2004). The human sense of smell: Are we better than we think? *Public Library of Science, Biology, 2,* 572–575.

62 But there's no doubt that your most powerful: Shepherd (2004), The human sense of smell.

62 This is especially true of odors: Zelano, C., & Sobel, N. (2005). Humans as an animal model for systems-level organization of olfaction. *Neuron, 48,* 431–454.

62 It turns out that 50 percent of the population: Zelano & Sobel (2005), Humans as an animal model.

63 your nose literally changes: Wang, L., Chen, L., & Jacob, T. (2004). Evidence for peripheral plasticity in human odour response. *Journal of Physiology (London), 554,* 236–244.

63 In fact, scientists have found very few: Zelano & Sobel (2005), Humans as an animal model.

63 In one study, two closely related: Li, W., Howard, J. D., Parrish, T. B., & Gottfried, J. A. (2008). Aversive learning enhances perceptual and cortical discrimination of indiscriminable odor cues. *Science, 319,* 1842–1845.

65 For example, if you were exposed to high concentrations: Li, W., Luxenberg, E., Parrish, T., and Gottfried, J. A. (2006). Learning to smell the roses: Experience-dependent neural plasticity in human piriform and orbitofrontal cortices. *Neuron, 52,* 1097–1108.

66 JC Ho notwithstanding, most of us are not particularly skilled: Stevenson, R. J., & Case, T. I. (2005). Olfactory imagery: A review. *Psychonomic Bulletin and Review, 12,* 244–264.

66 In a particularly entertaining example: Morrot, G., Brochet, F., & Dubourdieu, D. (2001). The color of odors. *Brain and Language, 79,* 309–320.

66 It could be that our failings to both explicitly name: Stevenson & Case (2005), Olfactory imagery.

67 He also learned that anosmia can be notoriously difficult: Cowart, B. J., Young, I. M., Feldman, R. S., & Lowry, L. D. (1997). Clinical disorders of smell and taste. *Occupational Medicine, 12,* 465–483.

69 A few days after your birth, you could recognize: Zelano & Sobel (2005), Humans as an animal model.

69 newborns show a preference for the odor: Schaal, B., Marlier, L., & Soussignan, R. (1998). Olfactory function in the human fetus: Evidence from selective neonatal responsiveness to the odor of amniotic fluid. *Behavioral Neuroscience, 112,* 1438–1449.

69 By the time you were five, you were able to identify: Weisfeld, G. E., Czilli, T., Phillips, K. A., Gall, J. A., & Lichtman, C. M. (2003). Possible olfaction-based mechanisms in human kin recognition and inbreeding avoidance. *Journal of Experimental Child Psychology, 85,* 279–295.

69 And if you have children, you could easily recognize the odor: Russell, M. J., Mendelson, T., & Peeke, V. S. (1983). Mothers' identification of their infant's odors. *Ethological Sociobiology, 4,* 29–31.

69 The one interesting exception is if your children: Weisfeld et al. (2003), Possible olfaction-based mechanisms in human kin.

69 In fact, parents have much less difficulty: Weisfeld et al. (2003), Possible olfaction-based mechanisms in human kin.

69 Strangers also have an easier time: Weisfeld et al. (2003), Possible olfaction-based mechanisms in human kin.

69 And even scent-tracking dogs: Harvey, L. M., Harvey, S. J., Hom, M., & Perna, A. (2006). The use of bloodhounds in determining the impact of genetics and the environment on the expression of human odortype. *Journal of Forensic Science, 51,* 1109–1114.

69 While you would have no trouble: Weisfeld et al. (2003), Possible olfaction-based mechanisms in human kin.

69 It could very well be that you recognize your kin's odors: Weisfeld et al. (2003), Possible olfaction-based mechanisms in human kin.

70 Research suggests that you have the ability to recognize: Weisfeld et al. (2003), Possible olfaction-based mechanisms in human kin.

70 This influence likely comes from a particular set of genes: Singh, P. B. (2001). Chemosensation and genetic individuality. *Reproduction, 121,* 529–539.

70 For instance, a woman's scent will change: Kuukasjarvi, S., Eriksson, C.J.P., Koskela, E., Mappes, T., Nissinen, K., & Rantala, M. J. (2004). Attractiveness of women's body odors over the menstrual cycle: The role of oral contraceptives and receiver sex. *Behavioral Ecology, 15,* 579–584.

71 Mood can also influence one's odor: Ackerl, K., Atzmueller, M., & Grammer, K. (2002). The scent of fear. *Neuroendocrinology Letters, 23,* 79–84.

71 And recent research shows that simply eating meat: Havlicek, J., & Lenochova, P. (2006). The effect of meat consumption on body odor attractiveness. *Chemical Senses, 31,* 747–752.

71 In a recent study, mothers were asked to rate: Case, T. I., Repacholi, B. M., & Stevenson, R. J. (2006). My baby doesn't smell as bad as yours: The plasticity of disgust. *Evolution and Human Behavior, 27,* 357–365.

72 Research shows that dog owners can nearly always: Wells, D. L., & Hepper, P.G. (2000). The discrimination of dog odours by humans. *Perception, 29,* 111–115.

72 Imagine yourself in this study: Herz, R. S. (2004). A naturalistic analysis of autobiographical memories triggered by olfactory visual and auditory stimuli. *Chemical Senses, 29,* 217–224.

73 In a study also conducted by Herz and her colleagues: Herz, R. S., Eliassen, J., Beland, S., & Souza, T. (2004). Neuroimaging evidence for the emotional potency of odor-evoked memory. *Neuropsychologia, 42,* 371–378.

74 There is substantial evidence that emotional memories: Wells, G. L., & Loftus, E. F. (2003). Eyewitness memory for people and events. In Goldstein, A. M. (Ed.), *Handbook of Psychology: Forensic Psychology* (Vol. 11, pp. 149–160). New York: John Wiley.

74 Thus, in being able to functionally render: Herz (2004), A naturalistic analysis.

75 There is evidence that when asked to determine: Sulmont, C., Issanchou, S., & Koster, E. P. (2002). Selection of odorants for memory tests on the basis of familiarity, perceived complexity, pleasantness, similarity and identification. *Chemical Senses, 27,* 307–317.

75 Brain imaging shows that separate brain regions: Royet, J. P., Zald, D., Versace, R., Costes, N., Lavenne, F., Koenig, O., & Gervais, R. (2000). Emotional responses to pleasant and unpleasant olfactory, visual, and auditory stimuli: A positron emission tomography study. *Journal of Neuroscience, 20,* 7752–7759.

75 And three-day-old infants show reactions: Soussignan, R., Schaal., B., Marlier, L., & Jiang, T. (1997). Facial and autonomic responses to biological and artificial olfactory stimuli in human neonates: Reexamining early hedonic discrimination of odors. *Physiology and Behavior, 4,* 745–758.

75 The approach/avoidance distinction: Herz, R. S., & Engen, T. (1996). Odor memory: Review and analysis. *Psychonomic Bulletin & Review, 3,* 300–313.

75 But as the Masai tribe in Kenya: Herz, R. S. (2007). *The scent of desire: Discovering our enigmatic sense of smell.* New York: William Morrow/HarperCollins Publishers.

76 So, while Americans tend to like the smell of wintergreen: Cain, W. S., & Johnson, F. Jr. (1978). Lability of odor pleasantness: Influence of mere exposure. *Perception, 7,* 459–465.

76 You would likely rate an odor contained in a bottle: Herz, R. S., & von Clef, J. (2001).

The influence of verbal labeling on the perception of odors: Evidence for olfactory illusions? *Perception, 30,* 381–391.

76 Herz also found that a subject's opinion: Herz (2007), *The scent of desire.*

76 And this malleability of odor preferences: Herz (2007), *The scent of desire.*

76 One study used data from evaluations: Khan, R. M., Luk, C. H., Flinker, A., Aggarwal, A., Lapid, H., Haddad, R., & Sobel, N. (2007). Predicting odor pleasantness from odorant structure: Pleasantness as a reflection of the physical world. *The Journal of Neuroscience, 27,* 10015–10023.

CHAPTER 4. *Like Marvin Gaye for Your Nose*

79 It was a risky proposition, costing: Binkley, C. (2008). *Winner takes all: Steve Wynn, Kirk Kerkorian, Gary Loveman, and the race to own Las Vegas.* New York: Hyperion.

80 The Moors developed a method: Handel, N. (2003). The sweet smell of excess. *The Los Angeles Times Magazine,* June 15.

80 Bloomingdale's currently uses a different scent: Ravn, K. (2007). Smells like sales. *The Los Angeles Times,* August 20.

81 Marketing researchers have shown that people: Holland, R. W., Hendriks, M., & Aarts, H. (2005). Smells like clean spirit. Nonconscious effects of scent on cognition and behavior. *Psychological Science, 16,* 689–693.

81 Other studies show that patrons: Holland et al. (2005), Smells like clean spirit.

81 But the research also shows that these same effects: Spangenberg, E., Crowley, A., & Henderson, P. W. (1996). Improving the store environment: The impact of ambient scent on evaluations of and behaviors in a store. *Journal of Marketing, 60,* 67–80.

82 Your nose is likely using odors all of the time: Hummel, T., Mojet, J., & Kobel, G. (2006). Electro-olfactograms are present when odorous stimuli have not been perceived. *Neuroscience Letters, 397,* 224–228.

82 Despite the fact that they are too weak: Hummel et al. (2006), Electro-olfactograms are present.

83 In fact, by the time you notice a smell: Pause, B. M. (2002). Human brain activity during the first second after odor presentation. In C. Rouby, B. Schaal, D. Dubois, R. Gervais, & A. Holley (Eds.), *Olfaction, taste and cognition* (pp. 309–323). Cambridge, U.K.: Cambridge University Press.

83 Imagine sitting in a room for 45 minutes: Degel, J., & Köster, E. P. (1999). Odors: implicit memory and performance effects. *Chemical Senses, 24,* 317–325.

83 For example, you would more quickly recognize: Holland et al. (2005), Smells like clean spirit.

84 It turns out that you sniff in your sleep: Stuck, B. A. (2008). The impact of olfactory stimulation on dreams. Presentation at the 2008 meeting of the American Academy of Otolaryngology and Head and Neck Surgery, September 21.

84 Sniffing in your sleep can also influence: Rasch, B., Buchel, C., Gais, S., Rasch, J. B. (2007). Odor cues during slow-wave sleep prompt declarative memory consolidation. *Science, 315,* 1426–1429.

84 Research shows that the presence of an undetectable: Li, W., Moallem, I., Paller, K. A., & Gottfried, J. A. (2007). Subliminal smells can guide social preferences. *Psychological Science, 18,* 1044–1049.

84 Research has shown that subjects are significantly faster: Platek, S. M., Thomson, J. W., & Gallup, G. G., Jr. (2004). Cross-modal self recognition: The role of visual, auditory, and olfactory primes. *Consciousness and Cognition, 13,* 197–210.

85 If you were exposed to low levels of odor samples: Pause, B. M., Ohrt, A., Prehn, A., & Ferstl, R. (2004). Positive emotional priming of facial affect perception in females is diminished by chemosensory anxiety signals. *Chemical Senses, 29,* 797–805.

85 It turns out that the startle reflex: Prehn, A., Ohrt, A., Sojka, B., Ferstl, R., & Pause, B. M. (2006). Chemosensory anxiety signals augment the startle reflex in humans. *Neuroscience Letters, 394,* 127–130.

85 For example, when subjects are asked to: Chen, D., Katdare, A., & Lucas, N. (2006). Chemosignals of fear enhance cognitive performance in humans. *Chemical Senses, 31,* 415–423.

85 Animals ranging from amoeba to deer: Chen et al. (2006), Chemosignals of fear enhance cognitive performance.

87 In her original 1971 report of the study: McClintock, M. K. (1971). Menstrual synchrony and suppression. *Nature, 229,* 244–245.

87 Regarding the former, it turns out that self-reports: Schank, J. C. (2006). Do human menstrual-cycle pheromones exist? *Human Nature, 17,* 448–470.

87 Next, a replication that did find menstrual entrainment: Schank, Do human menstrual-cycle pheromones exist?

88 In 1998 McClintock and her colleague Kathleen Stern: Stern, K., & McClintock, M. K. (1998). Regulation of ovulation by human pheromones. *Nature, 392,* 177–179.

88 Pheromones in the animal world are typically considered: Wysocki, C. J., & Preti, G. (2004). Facts, fallacies, fears, and frustrations with human pheromones. *The Anatomical Record Part A, 281,* 1201–1211.

88 In contrast, *primer pheromones* are the chemical signals: Wysocki & Preti (2004), Facts, fallacies, fears, and frustrations with human pheromones.

89 There is evidence, for example, that women who live with: Veith, J. L., Buck. M., Getzlaf, S., Van Dalfesen, P., & Slade, S. (1983). Exposure to men influences the occurrence of ovulation in women. *Psychology and Behavior, 31,* 313.

89 A related study has shown that regularly applying: Cutler, W. B., Preti, G., & Krieger, A. (1986). Human axillary secretions influence women's menstrual cycles: The role of donor extract from men. *Hormones and Behavior, 20,* 463–473.

89 Typically the criticism centers on the number: Hays, W.S.T. (2003). Human pheromones: Have they been demonstrated? *Behavioral Ecology and Sociobiology, 54,* 89–97.

89 And there have been claims of pheromone influences: McCoy, N. L., & Pitino, L. (2002). Pheromonal influences on sociosexual behavior in young women. *Physiology & Behavior, 75,* 367–375.

89 However, these claims have been strongly criticized: Hays (2003), Human pheromones.

90 It turns out that if a *three-day-old* infant is placed on: Varendi, H., & Porter, R. H. (2001). Breast odour as the only maternal stimulus elicits crawling towards the odour source. *Acta Paediatrica, 90,* 372–375.

91 And McClintock along with many other human pheromone researchers: Wysocki & Preti (2004), Facts, fallacies, fears, and frustrations with human pheromones.

91 The modified definition of pheromones proffered: Jacob, S., & McClintock, M. K. (2000). Psychological state and mood effects of steroidal chemosignals in women and men. *Hormones & Behavior, 37,* 57–78.

91 Thus, the more dissimilar the MHC code between your parents: Penn, D. J. (2002). The scent of genetic compatibility: Sexual selection and the major histocompatibility complex. *Ethology, 108,* 1–21.

92 Animals as diverse as mice, birds, fish: Penn (2002), The scent of genetic compatibility.

92 A number of T-shirt tests have found that both men: Wedekind, C., & Furi, S.

(1997). Body odour preferences in men and women: Do they aim for specific MHC combinations or simply heterozygosity? *Proceedings of the Royal Society, Ser. B, 264,* 1471–1479.

92 There is also some evidence that married couples: Ober, C., Weitkamp, L. R., Cox, N., Dytch, H., Kostyu, D., & Elias, S. (1997). HLA and mate choice in humans. *The American Journal of Human Genetics, 61,* 497–504.

92 Interestingly, for women, sensitivity to MHC compatibility: Thornhill, R., Gangestad, S. W., Miller, R. D., Scheyd, G. J., McCollough, J. K., & Franklin, M. (2003). Major histocompatibility complex genes, symmetry, and body scent attractiveness in men and women. *Behavioral Ecology, 14,* 668–678.

92 There is also recent evidence suggesting that: Garver-Apgar, C. E., Gangestad, S. W., Thornhill, R., Miller, R. D., Olp, J. J. (2006). Major histocompatibility complex alleles, sexual responsivity, and unfaithfulness in romantic couples. *Psychological Science, 10,* 830–835.

93 When women are in their fertile periods: Thornhill, R., & Gangestad, S. W. (1999). The scent of symmetry: A human sex pheromone that signals fitness? *Evolution and Human Behavior, 20,* 175–201.

93 It's long been known that low body symmetry: Thornhill & Gangestad (1999), The scent of symmetry.

93 Low symmetry is predictive of worse: Thornhill et al. (2003), Major histocompatibility complex genes, symmetry, and body scent.

93 In contrast, research shows that men with *high* symmetry: Thornhill et al. (2003), Major histocompatibility complex genes, symmetry, and body scent.

93 For a woman in her fertile phase, the degree: Thornhill & Gangestad (1991), The scent of symmetry.

94 Women in their fertile phase rate the odors: Havlíček, J., Roberts, S. C., & Flegr, J. (2005). Women's preference for dominant male odour: Effects of menstrual cycle and relationship status. *Biology Letters, 1,* 256–259.

94 Three recent studies have shown that men detect: Havlíček, J., Dvořáková, R., Bartoš, L., & Flegr, J. (2006). Non-advertized does not mean concealed: Body odour changes across the human menstrual cycle. *Ethology, 112,* 81–90; Kuukasiarvi, S., Eriksson, C.J.P., Koskela, E., Mappes, T., Nissinen, K., & Rantala, M. J. (2004). Attractiveness of women's body odors over the menstrual cycle: The role of oral contraceptives and receiver sex. *Behavioral Ecology, 15,* 579–584; Singh, D., & Bronstad, P. M. (2001). Female body odour is a potential cue to ovulation. *Proceedings of the Royal Society of London, Ser. B, 268,* 797–801.

94 Chemical compounds related to putative pheromones: Monti-Bloch, L., & Grosser, B. (1991). Effect of putative pheromones on the electrical activity of the human vomeronasal organ. *Journal of Steroid Biochemistry and Molecular Biology, 39,* 573–582.

95 These same compounds can cause changes: Jacob, S., Kinnunen, L. H., Metz, J., Cooper, M., & McClintock, M. K. (2001). Sustained human chemosignal unconsciously alters brain function. *NeuroReport, 12,* 2391–2394.

95 In a particularly compelling example: McClintock, M. K. (2002). The neuroendocrinology of social chemosignals in humans and animals: Odors, pheromones and vasanas. In D. Pfaff, A. Arnold, A. Etgen, R. Rubin, & S. Fahrbach (Eds.), *Hormones, brain & behavior* (pp. 797–870). San Diego, CA: Academic Press.

95 In another study conducted by Martha McClintock: Jacob, S., & McClintock, M. K. (2000). Psychological state and mood effects of steroidal chemosignals in women and men. *Hormones and Behavior, 37,* 57–78.

95 Other research has shown that inhaling a male steroid: Wyart, C., Webster, W. W., Chen, J. H., Wilson, S. R., McClary, A., Khan, R. M., & Sobel, N. (2007). Smelling a single component of male sweat alters levels of cortisol in women. *Journal of Neuroscience, 27,* 1261–1265.

96 It turns out that the smells of nursing: McClintock, M. K., Bullivant, S., Jacob, S., Spencer, N., Zelano, B., & Ober, C. (2007). Human body scents: Conscious perceptions and biological effects. *Chemical Senses, 30* (suppl 1), i135–i137.

CHAPTER 5. *Cold Leftovers with a Fine North Dakota Cabernet*

103 Research in anthropology and evolutionary biology: Hkadik, C. M., Pasquet, P., & Simmen, B. (2002). New perspectives on taste and primate evolution: The dichotomy in gustatory coding for perception of beneficent versus noxious substances as supported by correlations among human thresholds. *American Journal of Physical Anthropology, 117,* 342–348.

105 For example, if you were presented with four glasses: Zampini, M., Sanabria, D., Phillips, N., & Spence, C. (2007). The multisensory perception of flavor: Assessing the influence of color cues on flavor discrimination responses. *Food Quality and Preference, 18,* 975–984; DuBose, C. N., Cardello, A. V., & Maller, O. (1980). Effects of colorants and flavorants on identification, perceived flavor intensity, and hedonic quality of fruit-flavored beverages and cake. *Journal of Food Science, 45,* 1393–1399.

105 For example, if a weak drink were given a bright color: DuBose et al. (1980), Effects of colorants and flavorants on identification.

105 Lest you think these results are specific: Delwiche, J. (2004). The impact of perceptual interactions on perceived flavor. *Food Quality and Preference, 15,* 137–146.

106 Visible texture can influence how creamy: de Wijk, R. A., Polet, I. A., Engelen, L., van Doorn, R. M., & Prinz, J. F. (2004). Amount of ingested custard dessert as affected by its color, odor, and texture. *Physiology & Behavior, 82*(2–3 L1), 397–403.

106 Appearance can also influence your ability: Delwiche (2004), The impact of perceptual interactions on perceived flavor.

106 Finally, the influences of vision on flavor: Delwiche (2004), The impact of perceptual interactions on perceived flavor.

106 Research on primates shows that the same cell groups: Rolls, E. T., & Baylis, L. L. (1994). Gustatory, olfactory, and visual convergence within the primate orbito-frontal cortex. *Journal of Neuroscience, 14,* 5437–5452.

106 Related research shows that seeing food: Stillman, J. A. (2002). Gustation: Intersensory experience par excellence. *Perception, 31,* 1491–1500.

106 Brain scanning studies on humans: Royet, J. P., Koenig, O., Gregoire, M-C., Cinotti, L., Lavenne, F., Le Bars, D., Costes, N., et al. (1999). Functional anatomy of perceptual and semantic processing for odors. *Journal of Cognitive Neuroscience, 11,* 94–109.

106 Research shows that the sounds we hear: Christensen, C. M., & Vickers, Z. M. (1981). Relationship of chewing sounds to judgments of food crispness. *Journal of Food Science, 46,* 574–578.

106 These sounds reach your inner ears through two routes: Zampini, M., & Spence, C. (2004). The role of auditory cues in modulating the perceived crispness and staleness of potato chips. *Journal of Sensory Studies, 19,* 347–364.

107 In an experiment that won the coveted Ig Nobel Prize: Zampini & Spence (2004), The role of auditory cues.

110 This *tactile capture* of flavor is actually very common: Delwiche, J. F., Lera, M. F., & Breslin, A. S. (2000). Selective removal of a target stimulus localized by taste in humans. *Chemical Senses, 25,* 181–187.

111 But your sense of a food's texture or viscosity: Delwiche (2004), The impact of perceptual interactions on perceived flavor.

111 Research shows that your sensitivity to tastes is at its peak: Delwiche (2004), The impact of perceptual interactions on perceived flavor.

112 It turns out that temperature itself can change: Cruz, A., & Green, B. G. (2000). Thermal stimulation of taste. *Nature, 403,* 889–892.

112 These irritants affect the sensitive touch fibers: Verhagen, J. V., & Engelen, L. (2006). The neurocognitive bases of human multimodal food perception: Sensory integration. *Neuroscience and Biobehavioral Reviews, 30,* 613–650.

113 Research shows that one of the most common: Verhagen & Engelen (2006), The neurocognitive bases of human multimodal food perception.

113 Like all pain, though, the best part is when it stops: Verhagen & Engelen (2006), The neurocognitive bases of human multimodal food perception.

114 Smell mostly affects taste not from your inhaling: Shepherd, G. M. (2006). Smell images and the flavour system in the human brain. *Nature, 444,* 316–321.

114 For example, if you were asked to chew a flavorless gum: Stillman (2002), Gustation.

114 The intensity of a presented odor: Stillman (2002), Gustation.

114 And just as a sweet flavor can suppress the strength: Stillman (2002), Gustation.

114 In fact, when you see an ingredient listed: Stevenson, R. J., & Tomiszek, C. (2007). Olfactory-induced synesthesias: A review and model. *Psychological Bulletin, 133,* 294–309.

114 This is consistent with neurophysiological research: Verhagen & Engelen (2006), The neurocognitive bases of human multimodal food perception.

115 Brain scanning shows that when a consistent: Verhagen & Engelen (2006), The neurocognitive bases of human multimodal food perception.

115 If you were presented with a very low-level cherry: Dalton, P., Doolittle, N., Nagata, H., & Breslin, P.A.S. (2000). The merging of the senses: Integration of subthreshold taste and smell. *Nature Neuroscience, 3,* 431–432.

115 If you were asked to taste two solutions: Djordjevic, J., Zatorre, R. G., & Jones-Gotman, M. (2004). Effects of Perceived and Imagined Odors on Taste Detection. *Chemical Senses, 29,* 199–208.

115 And it turns out that not only do all of your senses: Verhagen & Engelen (2006), The neurocognitive bases of human multimodal food perception.

116 Imagine being in this experiment: Yeomans, M. R., Chambers, L., Blumenthal, H., & Blake, A. (2008). The role of expectancy in sensory and hedonic evaluation: The case of smoked salmon ice-cream. *Food Quality and Preference, 19,* 565–573.

116 For example, if you were poured a glass of wine: Wansing, B., Payne, C. R., North, J. (2007). Fine as North Dakota wine: Sensory expectations and the intake of companion foods. *Physiology & Behavior, 90,* 712–716.

117 The effect works in an actual restaurant: Wansing et al. (2007), Fine as North Dakota wine.

117 Brain imaging research shows that as subjects sip: Plassman, H., O'Doherty, J., Shiv, B., & Rangel, A. (2008). Marketing actions can modulate neural representations of experienced pleasantness. *Proceedings of the National Academy of Sciences, USA, 105,* 1050–1054.

119 But research shows that wine experts do not: Parr, W. V., Heatherbell, D., & White, K. G. (2002). Demystifying wine expertise: Olfactory threshold, perceptual skill and semantic memory in expert and novice wine judges. *Chemical Senses, 27,* 747–755.

119 Furthermore, novices seem to be nearly as good: Parr et al. (2002), Demystifying wine expertise.

119 Recent brain scans of sommeliers reveal: Castriota-Scanderberg, A., Hagberg, G. E., Cerasa, A., Committeri, G., Galati, G., Patria, F., Pitzalis, S., Caltagirone, C., & Franckowiak, R. (2005). The appreciation of wine by sommeliers: A functional magnetic resonance study of sensory integration. *NeuroImage* 25, 570–578.

119 Wine experts have developed a rich conceptual: Hughson, A. L., & Boakes, R. A. (2001). Perceptual and cognitive aspects of wine expertise. *Australian Journal of Psychology,* 53, 103–108.

120 Thus, experts almost always show superior skills: Hughson & Boakes (2001), Perceptual and cognitive aspects of wine expertise.

120 In fact, this conceptual advantage is shared: Ericsson, K. A., & Lehmann, A. C. (1996). Expert and exceptional performance: Evidence of maximal adaptation to task constraints. *Annual Review of Psychology,* 47, 273–305.

121 In memory experiments, wine experts are better able: Hughson, A. L., & Boakes, R. A. (2002). The knowing nose: The role of knowledge in wine expertise. *Food Quality and Preference, 13,* 463–472.

121 This phenomenon, known as *verbal overshadowing:* Melcher, J. M., & Schooler, J. W. (1996). The misremembrance of wines past: Verbal and perceptual expertise differentially mediate verbal overshadowing of taste memory. *Journal of Memory and Language,* 35, 231–245.

122 Regular wine drinkers (two glasses a week): Melcher & Schooler (1996), The misremembrance of wines past.

122 When experts are asked to recognize: Melcher & Schooler (1996), The misremembrance of wines past.

CHAPTER 6. *Rubber Hands and Rubber Brains*

130 But in fact, only about 10 percent: *National Federation of Blind Report.* March 26, 2009. http://www.nfb.org/nfb/Default.asp, retrieved March 25, 2009.

130 But even blind individuals who have no Braille: Goldreich, D., & Kanics, I. M. (2003). Tactile acuity is enhanced in blindness. *The Journal of Neuroscience,* 23, 3439–3445.

130 In fact, a blind individual's touch sensitivity: Goldreich & Kanics (2003), Tacticle acuity is enhanced in blindness.

130 And for blind individuals, brain areas: Amedi, A., Merabet, L. B., Bermpohl, F., & Pascual-Leone, A. (2005). The occipital cortex in the blind: Lessons about plasticity and vision. *Current Directions in Psychological Science, 14,* 306–311.

131 There are numerous reports of experienced Braille: Amedi et al. (2005), The occipital cortex in the blind.

131 When using this device on the touch strip: Pascual-Leone, A., Walsh, V., & Rothwell, J. (2000). Transcranial magnetic stimulation in cognitive neuroscience—virtual lesion, chronometry, and functional connectivity. *Current Opinion in Neurobiology, 10,* 232–237.

131 Imagine yourself in this experiment: Pascual-Leone, A., & Hamilton, R. (2001). The metamodal organization of the brain. In C. Casanova & M. Ptito (Eds.) *Progress in brain research* (Vol. 134, pp. 1–19).

132 Five days of blindfolding would change you: Pascual-Leone & Hamilton (2001), The metamodal organization of the brain.

133 That's all it takes to enlist visual brain: Facchini, S., & Aglioti, S. M. (2003). Short- term light deprivation increases tactile spatial acuity in humans. *Neurology, 60,* 1998–1999; Merabet, L. B., Swisher, J. D., McMains, S. A., Halko, M. A., Amedi, A.,

Pascual-Leone, A., & Somers, D. C. (2007). Combined activation and deactivation of visual cortex during tactile sensory processing. *Journal of Neurophysiology, 97,* 1633–1641.

133 Ninety minutes in the dark are also enough: Weisser, V., Stilla, R., Peltier, S., Hu, X., & Sathian, K. (2005). Short-term visual deprivation alters neural processing of tactile form. *Experimental Brain Research, 166,* 572–582.

134 Even after the five days of darkness: Pascual-Leone & Hamilton (2001), The metamodal organization of the brain.

134 But recent experiments using brain scanning: Sathian, K., & Zangaladse, A. (2002). Feeling with the mind's eye: Contribution of visual cortex to tactile perception. *Behavioural Brain Research, 135,* 127–132.

135 This device is the Tongue Display Unit, TDU: Bach-y-Rita, P. W., & Kercel, S. (2003). Sensory substitution and the human-machine interface. *Trends in Cognitive Science, 7,* 541–546.

135 Without any training, subjects would be able: Bach-y-Rita & Kercel (2003), Sensory substitution and the human-machine interface.

135 But after many of hour of training: Bach-y-Rita & Kercel (2003), Sensory substitution and the human-machine interface.

135 Most recently, touch-based vision substitution systems: Bach-y-Rita & Kercel (2003), Sensory substitution and the human-machine interface.

136 After a few minutes, you can recognize shapes: Bach-y-Rita, P., Kaczmarek, K. A., Tyler, M. E., & Garcia-Lara, J. (1998). Form perception with a 49-point electrotactile stimulus array on the tongue: A technical note. *The Journal of Rehabilitation Research and Development, 35,* 427–430.

137 It turns out that the amount of touch strip: Sterr, A., Muller, M. M., Elbert, T., Rockstroh, B., Pantev, C., & Taub, E. (1998). Changed perceptions in Braille readers. *Nature, 391,*134–135.

137 For violin players, the touch brain's mapped space: Elbert, T., Pantev, C., Wienbruch, C., Rockstroh, B., & Taub, E. (1995). Increased cortical representation of the fingers of the left hand in string players. *Science, 270,* 305–307.

137 Professional pianists have much better "two-point": Ragert, P., Schmidt, A., Altenmuller, E., & Dinse, H. R. (2004). Superior tactile performance and learning in professional pianists: Evidence for meta-plasticity in musicians. *European Journal of Neuroscience, 19,* 473–478.

137 Imagine being given a set of tongs: Schaefer, C. A., Rothemund, Y., Jeinze, H. J., & Rotte, M. (2004). Short-term plasticity of the primary somatosensory cortex during tool use. *NeuroReport, 15,* 1293–1297.

138 In the touch stimulation method known as *coactivation*: Hodzic, A., Veit, R., Karim, A. A., Erb, M., & Godde, B. (2004). Improvement and decline in tactile discrimination behavior after cortical plasticity induced by passive tactile coactivation. *Journal of Neuroscience, 24,* 442–446.

138 With three hours of coactivation: Hodzic et al. (2004), Improvement and decline in tactile discrimination behavior.

138 In turn, your two-point sensitivity improves: Hodzic et al. (2004), Improvement and decline in tactile discrimination behavior.

138 One study showed that after three hours: Dinse, H. R., Kleibel, N., Kalisch, T., Ragert, P., Wilimzig, C., & Tegenthoff, M. (2006). Tactile coactivation resets age-related decline of human tactile discrimination. *Annals of Neurology, 60,* 88–94.

138 Brain scanning of these subjects: Dinse, H. R. (2006). Cortical reorganization in the aging brain. In A. R. Møller (Ed.), *Progress in brain research* (Vol. 157, pp. 57–82).

139 With three hours of coactivation stimulation: Godde, B., Stauffenberg, B., Spengler,

F. & Dinse, H. R. (2000). Tactile coactivation-induced changes in spatial discrimination performance. *Journal of Neuroscience, 20,* 1597–1604.

139 This *short-term* plasticity is a result: Amedi et al. (2005), The occipital cortex in the blind.

140 This long-term plasticity is thought: Amedi et al. (2005), The occipital cortex in the blind.

140 A group of nonmusicians were taught to use their: Pascual-Leone, A. (1996). Reorganization of cortical motor outputs in the acquisition of new motor skills. In J. Kinura & H. Shibasaki (Eds.), *Recent advances in clinical neurophysiology* (pp. 304–308). Amsterdam: Elsevier Science; Pascual-Leone, A., Nguyet, D., Cohen, L. G., Brasil-Neto, J. P., Cammarota, A., & Hallett, M. (1995). Modulation of muscle responses evoked by transcranial magnetic stimulation during the acquisition of new fine motor skills. *Journal of Neurophysiology, 74,* 1037–1045.

142 For decades phantom limb pain: Flor, H., Nikolajsen, L., & Jensen, T. S. (2006). Phantom limb pain: A case of maladaptive CNS plasticity? *Nature Reviews Neuroscience, 7,* 873–881.

142 In fact, individuals with constant and intense phantom: Ramachandran, V. S., & Hirstein, W. (1998). The perception of phantom limbs: The D. O. Hebb lecture. *Brain, 121,* 1603–1630.

143 Newer options making use of neuroplasticity: Flor et al. (2006), Phantom limb pain.

143 In one treatment, mirrors are positioned: Ramachandran & Hirstein (1998), The perception of phantom limbs.

144 Recall the experiment in which subjects were taught: Pascual-Leone (1996), Reorganization of cortical motor outputs; Pascual-Leone et al. (1995), Modulation of muscle responses evoked by transcranial magnetic stimulation.

144 Most astonishing, however, were the results of the brain scans: Pascual-Leone (1996), Reorganization of cortical motor outputs; Pascual-Leone et al. (1995), Modulation of muscle responses evoked by transcranial magnetic stimulation.

145 It turns out that experienced Tai Chi practitioners: Kerr, C. E., Shaw, J. R., Wasserman, R. H., Chen, V. W., Kanojia, A., Bayer, T., & Kelley, J. M. (2008). Tactile acuity in experienced Tai Chi practitioners: Evidence for use dependent plasticity as an effect of sensory-attentional training. *Experimental Brain Research, 188,* 317–322.

145 Research shows that concentrating on different aspects: Braun, C., Haug, M., Wiech, K., Birbaumer, N., Elbert, T., & Roberts, L. (2002). Functional organization of primary somatosensory cortex depends on the focus of attention. *NeuroImage, 17,* 1451–1458.

145 To truly appreciate the oddity of these effects: Armel, K. C., & Ramachandran, V. S. (2003). Projecting sensations to external objects: Evidence from skin conductance response. *Proceedings of the Royal Society of London, Ser. B, 270,* 1499–1506.

146 If the lights were then turned off: Tsakiris, M., & Haggard, P. (2005). The rubber hand illusion revisited: Visuotactile integration and self-attribution. *Journal of Experimental Psychology: Human Perception and Performance, 31,* 80–91.

146 After you tell the experimenter you experience: Armel & Ramachandran (2003), Projecting sensations to external objects.

146 Or just when you least expect it: Lloyd, D., Morrison, I., & Roberts, N. (2006). Role for human posterior parietal cortex in visual processing of aversive objects in peripersonal space. *Neurophysiology, 95,* 205–214.

146 And your brain would react as if: Ehrsson, H. H., Wiech, K., Weiskopf, N., Dolan, R. J., & Passingham, R. E. (2007) Threatening a rubber hand that you feel is yours elicits a cortical anxiety response. *Proceedings of the National Academy of Sciences USA, 104,* 9828–9833.

146 After 45 minutes of watching your rubber hand: Schaefer, M., Flor, H., Heinze, H-J., & Rotte, M. (2006). Dynamic modulation of the primary somatosensory cortex during seeing and feeling a touched hand. *NeuroImage, 15,* 585–587.

146 Imagine looking down at your rubber hand: Schaefer, M., Flor, H., Heinze, H-J., & Rotte, M. (2007). Morphing the body: Illusory feeling of an elongated arm affects somatosensory homunculus. *NeuroImage, 36,* 700–705.

147 In contrast, the opposite brain reorganization: Schaefer, M., Flor, H., Heinze, H-J., & Rotte, M. (2008). Observing the touched body magnified alters somatosensory homunculus. *NeuroReport, 19,* 901–905.

147 For you to feel that your two real hands: Moseley, G. L., Olthof, N., Venema, A., Don, S., Wijers, M., Gallace, A., & Spence, C. (2008). Psychologically induced cooling of a specific body part caused by the illusory ownership of an artificial counterpart. *Proceedings of the National Academy of Sciences, USA, 105,* 13169–13173.

147 To make matters worse, your poor, neglected hand: Moseley et al. (2008), Psychologically induced cooling of a specific body part.

148 You sit in a chair and wear video goggles: Ehrsson, H. H. (2007). The experimental induction of out-of-body experiences. *Science, 317,* 1048.

148 And in a rather sadistic confirmation of the effect: Ehrsson (2007), The experimental induction of out-of-body experiences.

CHAPTER 7. *Touching Speech and Feeling Rainbows*

151 The Tadoma method was taught widely: Reed, C. M., Rabinowitz, W. M., Durlach, N. I., Braids, L. D., Conway-Fithian, S., & Schultz, M. C. (1985). Research on the Tadoma method of speech communication. *Journal of the Acoustical Society of America, 77,* 247–257.

153 Generally the student is instructed to place a thumb vertically: Alcorn, S. K. (1932). The Tadoma method. *The Volta Review, 34,* 195–198.

155 Practiced Tadoma users are as adept: Reed et al. (1985), Research on the Tadoma method of speech communication.

155 Consonants that involve lip and jaw movements: Reed et al. (1985), Research on the Tadoma method of speech communication.

155 Research shows that with enough training: Reed et al. (1985), Research on the Tadoma method of speech communication.

155 Research shows that touching a talker's face: Gick, B., Johannsdottir, K. M., Gibraiel, D., Mulbauer, J. (2008). Tactile enhancement of auditory and visual speech perception in untrained perceivers. *Journal of the Acoustical Society of America, 123,* EL72–EL76.

156 And if you were asked to visually lip-read: Gick et al. (2008), Tactile enhancement of auditory and visual speech perception in untrained perceivers.

156 Imagine being in this experiment: Fowler, C. A., & Dekle, D. J. (1991). Listening with eye and hand: Cross-modal contributions to speech perception. *Journal of Experimental Psychology: Human Perception and Performance, 17,* 816–828.

156 While rarely experienced touched speech: Fowler & Dekle (1991), Listening with eye and hand.

157 Research suggests that you'd have little trouble: Casey, S. J., & Newell, F. N. (2005). The role of long-term and short-term familiarity in visual and haptic face recognition. *Experimental Brain Research, 166,* 583–591.

157 In fact, you need very little visual: Kilgour, A. R., & Lederman, S. J. (2002). Face recognition by hand. *Perception & Psychophysics, 64,* 339–352.
157 Imagine this experiment: Kilgour & Lederman (2002), Face recognition by hand.
157 You'd also be nearly as accurate recognizing a real: Lederman, S. J., Klatzky, R. L., Abramowicz, A., Salsman, K., Kitada, R., & Hamilton, C. (2007). Haptic recognition of static and dynamic expressions of emotion in the live face. *Psychological Science, 18,* 158–164.
158 It's well known that you are especially: Searcy, J. H., & Bartlett, J. C. (1996). Inversion and processing of component and spatial-relational information of faces. *Journal of Experimental Psychology: Human Perception and Performance, 22,* 904–915.
158 Turning a face upside down makes it: Kilgour, A. R., & Lederman, S. J. (2006). A haptic face-inversion effect. *Perception, 35,* 921–931.
159 In fact, unlike the control subjects: Kilgour, A. R., de Gelder, B., & Lederman, S. J. (2004). Haptic face recognition and prosopagnosia. *Neuropsychologia, 42,* 707–712.
159 But recent brain imaging research shows that: Kilgour, A., Kitada, R., Servos, P., James, T., & Lederman, S. J. (2005). Haptic face identification activates ventral occipital and temporal areas: An fMRI study. *Brain & Cognition, 59,* 246–257.
161 For example, with your eyes closed, you could grasp: Solomon, H. Y., & Turvey, M. T. (1988). Haptically perceiving the distances reachable with hand-held objects. *Journal of Experimental Psychology: Human Perception and Performance, 14,* 404–427.
161 If you instead grab the rod somewhere in the middle: Turvey, M. T. (1996). Dynamic touch. *American Psychologist, 51,* 1134–1152.
161 If the rod happens to have a significant bend: Turvey, M. T., Burton, G., Pagano, C., Solomon, H. Y., & Runeson, S. (1992). Role of the inertia tensor in perceiving object orientation by dynamic touch. *Journal of Experimental Psychology: Human Perception and Performance, 18,* 714–727.
162 You stand next to a table: Burton, G., Turvey, M. T., & Solomon, Y. (1990). Can shape be perceived by dynamic touch? *Perception & Psychophysics, 48,* 477–487.
162 When you lift something, your muscle-tendon: Turvey (1996), Dynamic touch.
163 Research suggests that you are superb at judging: Carello, C., Thuot, S., Anderson, K. L., & Turvey, M. T. (1999). Perceiving the sweet spot. *Perception, 28,* 307–320.
163 With simple wielding, both expert hockey players: Hove, P., Riley, M. A., & Shockley, K. (2006). Perceiving affordances of hockey sticks by dynamic touch. *Ecological Psychology, 18,* 163–189.
163 Research shows that you'd be able to judge: Wagman, J. B., & Taylor, K. R. (2004). Chosen striking location and the user–tool–environment system. *Journal of Experimental Psychology: Applied, 10,* 267–280.
163 You can perceive the lengths and bends of rods: Pagano, C. C., Fitzpatrick, P., & Turvey, M. T. (1993). Tensorial basis to the constancy of perceived object extent over variations of dynamic touch. *Perception & Psychophysics, 54,* 43–54.
163 And even using your knee or ankle: Hajnal, A., Fonseca, S. T., Harrison, S., Kinsella-Shaw, J. M., & Carello, C. (2007). Comparison of dynamic (effortful) touch by hand and foot. *Journal of Motor Behavior, 39,* 82–88.
164 On a more fine-grain scale, twiddling: Santana, M. V. (1999). Perceiving whole and partial extents of small objects by dynamic touch. *Ecological Psychology, 11,* 283–307.
167 And for many textures, ratings are comparable: Klatzky, R. L., & Lederman, S. J. (2002). Perceiving texture through a probe. In M. L. McLaughlin, J. P. Hespanha, & G. S. Sukhatme (Eds.), *Touch in virtual environments* (pp. 180–193). Upper Saddle River, NJ: Prentice Hall PTR.
167 Without practice you could accurately judge the height: Chan, T. C., & Turvey,

M. T. (1991). Perceiving the vertical distances of surfaces by means of a hand-held probe. *Journal of Experimental Psychology: Human Perception and Performance, 17,* 347–358.

167 Next, you would be similarly accurate: Carello, C., Fitzpatrick, P., & Turvey, M. T. (1992). Haptic probing: Perceiving the length of a probe and the distance of a surface probed. *Perception & Psychophysics, 51,* 580–598.

167 For example, you can swing a long wooden rod: Flascher, O., & Carello, C. (1990). Visual and haptic perception of passability for different styles of locomotion. Poster presented at the meeting of the Society for Ecological Psychology, Urbana Champaign, IL.

167 You can also use a long rod to determine: Burton, G. (1992). Nonvisual judgment of the crossability of path gaps. *Journal of Experimental Psychology: Human Perception and Performance, 18,* 698–713.

168 Finally, the research suggests you can use: Fitzpatrick, P., & Carello, C., Schmidt, R.C., & Corey, D. (1994). Haptic and visual perception of an affordance for upright posture. *Ecological Psychology, 6,* 265–287.

168 Whiskers help small animals navigate: Burton, G. (1993). Non-neural extensions of haptic sensitivity. *Ecological Psychology,* 5,105–124.

169 Research shows that humans can detect: Burton (1993), Non-neural extensions of haptic sensitivity.

169 There's even evidence that if you: Burton (1993), Non-neural extensions of haptic sensitivity.

170 In fact, some spiders construct their webs: Walcott, C. (1963). The effect of the web on vibration sensitivity in the spider *Acharanea tepidariorum* Koch. *Journal of Experimental Zoology, 141,* 191–244.

170 Imagine being in this experiment: Kinsella-Shaw, J. M., & Turvey, M. T. (1992). Haptic perception of object distance in a single-strand vibratory web. *Perception & Psychophysics, 52,* 625–638.

CHAPTER 8. *Facing the Uncanny Valley*

176 This concept was originally discussed by robot designer: Mori, M. (1970). The uncanny valley. *Energy,* 7(4), 33–35.

177 Watching a humanoid robot perform: Oztop, E., Franklin, D., Chaminade, T., Gordon, C. (2005). Human-humanoid interaction: Is a humanoid robot perceived as a human. *International Journal of Humanoid Robotics, 2,* 537–559.

178 Similarly, mirror neuron sites engage: Chaminade, T., Hodgins, J., Kawato, M. (2007). Anthropomorphism influences perception of computer-animated characters' actions. *SCAN, 2,* 206–216.

178 It turns out that the more real a computer-animated: MacDorman, K. F., Green, R. D., Ho, C.-C., & Koch, C. T. (2009). Too real for comfort? Uncanny responses to computer generated faces. *Computers in Human Behavior,* 25(3), 695–710.

178 This fact seems to be particularly true: Green, R. D., MacDorman, K. F., Ho, C.-C., & Vasudevan, S. K. (2008). Sensitivity to the proportions of faces that vary in human likeness. *Computers in Human Behavior, 24,* 2456–2474.

180 As a result, if Chinese subjects are shown an image: Ge, L., Luo, J., Nishimura, M., & Lee, J. (2003). The lasting impression of Chairman Mao: Hyperfidelity of familiar-face memory. *Perception, 32,* 601–614.

180 Recent research suggests that you can also detect: Bredart, S., & Devue, C. (2006). The accuracy of memory for faces of personally known individuals. *Perception, 35,* 101–106.

180 And you detect the presence of facial images: Palermo, R., & Rhodes, G. (2007).

Are you always on my mind? A review of how face perception and attention interact. *Neuropsychologia 45*, 75–92.

180 If you were asked to quickly read the name: Young, A. W., Ellis, A. W., Flude, B. M., McWeeny, K. H., & Hay, D. C. (1986). Face–name interference. *Journal of Experimental Psychology: Human Perception and Performance, 12*, 466–475.

180 A recent study shows that most of us: Windhager, S., Slice, D. E., Schaefer, K., Oberzaucher, E., Thorstensen, T., Grammer, K. (2008). Face to face: The perception of automotive designs. *Human Nature, 19*, 331–346.

180 Imagine being in this experiment: Stone, A., & Valentine, T. (2004). Better the devil you know? Nonconscious processing of identity and affect of famous faces. *Psychonomic Bulletin & Review, 11*, 469–474.

181 While prosopagnosia was long considered: Kennerknect, I., Grueter, T., Welling, B., Wetzek, S., Horst, J., Edwards, S., & Grueter, M. (2006). First report of prevalence of non-syndromic hereditary prosopagnosia (HPA). *American Journal of Medical Genetics, Part A, 140*, 1617–1622.

181 Despite having no conscious recognition: Young, A. W. (1998). *Face & mind*. Oxford: Oxford University Press.

181 They are also able to more quickly: Young (1998), *Face & mind*.

182 An hour after you were born: Johnson, M. H., Dziurawiec, S., Ellis, H. D., & Morton, J. (1991). Newborns preferential tracking of faces and its subsequent decline. *Cognition, 40*, 1–20.

182 By the time you were four days old: Walton, G. E., Bower, N.J.A., & Bower, T.G.R. (1992). Recognition of familiar faces by newborns. *Infant Behavior and Development, 15*, 265–269.

182 And when you were you two weeks old: Meltzoff, A. N. and Moore, M. K. (1977). Imitation of facial and manual gestures by human neonates. *Science, 198*, 75–78.

182 But your current facial expertise didn't: Zebrowitz, L. A. (2006). Finally, faces find favor. *Social Cognition, 24*, 657–701.

182 It turns out that to recognize a face: Carey, S., & Diamond, R. (1977). From piece-meal to configurational representation of faces. *Science, 195*, 312–314.

183 For example, your skills at recognizing miniscule: Ge et al. (2003), The lasting impression of Chairman Mao.

183 If, however, you saw this entire gruesome image: Thompson, P. (1980). Margaret Thatcher: A new illusion. *Perception, 9*, 483–484.

183 Your memory for a facial feature: Tanaka, J. W., & Farah, M. J. (1993). Parts and wholes in face recognition. *The Quarterly Journal of Experimental Psychology, 4*, 225–245.

183 It turns out that when these experts look at dog: Diamond, R., & Carey, S. (1986). Why faces are and are not special: An effect of expertise. *Journal of Experimental Psychology: General, 115*, 107–117.

183 The same is true of expert car designers: Diamond & Carey (1986), Why faces are not special.

183 Even naïve subjects intensively trained: Gauthier, I., and Tarr, M. J. (1997). Becoming a 'greeble' expert: Exploring mechanisms for face recognition. *Vision Research, 37*, 1673–1682

184 Imagine yourself in this experiment: Rosenblum, L. D., Smith, N. M., & Niehus, R. P. (2007). Look who's talking: Recognizing friends from visible articulation. *Perception, 36*, 157–159.

185 Using a variety of techniques, research has shown: Lander, K., & Chuang, L. (2005). Why are moving faces easier to recognize? *Visual Cognition, 12*, 429–442.

185 Four-month-old infants can discriminate faces: Spencer, J., O'Brien, J., Johnston,

A., & Hill, H. (2006). Infants' discrimination of faces by using biological motion cues. *Perception, 35,* 79–89.

185 And facial motion can also convey information: Rosenblum, L. D., Yakel, D. A., Baseer, N., Panchal, A., Nordarse, B. C., & Niehus, R. P. (2002). Visual speech information for face recognition. *Perception & Psychophysics, 64,* 220–229.

186 In a demonstration particularly relevant: Knappmeyer, B., Thornton, I. M., & Beulthoff, H. H. (2003). The use of facial motion and facial form during the processing of identity. *Vision Research, 43,* 1921–1936.

187 If, for example, you were presented with a subliminal: Myrphy, S.T., & Zajonc, R. B. (1993). Affect, cognition, and awareness: Affective priming with optimal and suboptimal stimulus exposures. *Journal of Personality and Social Psychology, 64,* 723–739.

187 If you were presented subliminal faces: Ruys, K. I., & Stapel, D. A. (2008). The secret life of emotions. *Psychological Science, 19,* 358–391.

187 Pretty likely—if, in fact, you were first presented: Ruys & Stapel (2008), The secret life of emotions.

188 When you see a smiling face, the muscles: Dimberg, U. (1982). Facial reactions to facial expressions. *Psychophysiology, 19,* 643–647.

188 Subliminal smiles induce subtle but measurable: Dimberg, U., Thunberg, M., & Elmehed, K. (2000). Unconscious facial reactions to emotional facial expressions. *Psychological Science, 11,* 86–89.

188 Research shows that if your covert reactions: Oberman, L. M., Winkielman, P., & Ramachandran, V. S. (2007). Face to face: Blocking facial mimicry can selectively impair recognition of emotional expressions. *Social Neuroscience, 2,* 167–178; Pitcher, D., Garrido, L., Walsh, V., & Duchaine, B. C. (2008). Transcranial magnetic stimulation disrupts the perception and embodiment of facial expressions. *The Journal of Neuroscience, 28,* 8929–8933.

189 If you were first put into a scared mood: Moody, E. J., McIntosh, D. N., Mann, L. J., & Weisser, K. R. (2007). More than mere mimicry? The influence of emotion on rapid facial reactions to faces. *Emotion, 7,* 447–457.

189 Hearing happy or fearful voices: Magnee, M. J., Stekelnburg, J. J., Kemner, C. Kemner, & de Gelder, B. (2007). Similar facial electromyographic responses to faces, voices, and body expressions. *NeuroReport, 18,* 369–372.

189 Imagine being in this experiment: Strack, F., Martin, L. L., & Stepper, S. (1988). Inhibiting and facilitating conditions of the human smile: A non-obtrusive test of the facial feedback hypothesis. *Journal of Personality and Social Psychology, 53,* 768–777.

191 Similar sneaky methods to induce: Niedenthal, P. M. (2007). Embodying emotion. *Science, 316,* 1002–1005.

191 One prominent explanation is that emotions: Niedenthal (2007), Embodying emotion.

191 Research suggests that by the time you're: Malatesta, C. Z., Fiore, M. J., & Messina, J. J. (1987). Affect, personality, and facial expressive characteristics of older people. *Psychology and Aging, 2,* 64–69.

192 This could explain why, as research has shown: Malatesta et al. (1987), Affect, personality, and facial expressive characteristics.

192 Research shows that you can establish an impression: Bar, M., Neta, M., & Linz, H. (2006). Very first impressions. *Emotion, 6,* 269–278.

193 In one experiment, a composite face: Little, A. C., & Perrett, D. I. (2007). Using composite images to assess accuracy in personality attribution to faces. *British Journal of Psychology, 98,* 111–126.

193 And in a very recent study: Boothroyd, L. G., Jones, B. C., Burt, M., Debruin, L. M.,

& Perrett, D. I. (2008). Facial correlates of sociosexuality. *Evolution and Human Behavior,* 3, 211–218.

194 It's long been known that people choose mates: Rushton, J. P., Russell, R.J.H., & Wells, P. A. (1985). Personality and genetic similarity theory. *Journal of Social and Biological Structures,* 8, 63–86.

194 But there is also evidence that spouses: Zajonc, R. B., Adelmann, P. K., Murphy, S. T., & Niendenthal, P. M. (1987). Convergence in the physical appearance of spouses. *Motivation and Emotion,* 11, 335–346.

196 Research has long shown that adults: Langlois, J. H., Kalakanis, L., Rubenstein, A. J., Larson, A., Hallam, M., & Smoot, M. (2000). Maxims or myths of beauty? A meta-analytic and theoretical review. *Psychological Bulletin,* 126, 390–423.

196 Over 30 separate studies: Langlois et al. (2000), Maxims or myths of beauty?

196 But there is evidence that in judging overall: Peters, M., Rhodes, G., & Simmons, L. W. (2007). Contributions of the face and body to overall attractiveness. *Animal Behaviour,* 73, 937–942.

196 When asked which candidates they would likely: Chiao, J. Y., Bowman, N. E., & Gill, H. (2008). The political gender gap: Gender bias in facial inferences that predict voting behavior. *The Public Library of Science ONE,* 3, e3666, 1–7.

196 Another recent experiment revealed that: Price, M. K. (2008). Fund-raising success and a solicitor's beauty capital: Do blondes raise more funds? *Economics Letters,* 100, 351–354.

196 Finally, analyses of contestant behavior: Belot, M., Bhaskar, V., & van de Ven, J. (2008). Beauty and the sources of discrimination. *ELSE Working Papers 241.* London, U.K.: ESRC Centre for Economic Learning and Social Evolution.

197 In fact, a recent study showing that more: Mobius, M., & Rosenblatt, T. (2006). Why beauty matters. *American Economic Review,* 96, 222–235.

197 When shown images of cropped faces: Langlois et al. (2000), Maxims or myths of beauty?

197 Based on ratings of these images, children: Langlois et al. (2000), Maxims or myths of beauty?

197 Adults with more attractive: Langlois et al. (2000), Maxims or myths of beauty?

197 For example, asked to rate the attractiveness: Olson, I. R., & Marshuetz, C. (2005). Facial attractiveness is appraised in a glance. *Emotion,* 5, 498–502.

198 And if a subliminal attractive face: Olson & Marshuetz (2005), Facial attractiveness is appraised in a glance.

198 In fact, there is a vast literature showing that many: Rhodes, G. (2006). The evolutionary psychology of facial beauty. *Annual Review of Psychology,* 57, 199–226.

198 This fact, along with its sociocultural power: Rhodes (2006), The evolutionary psychology of facial beauty.

198 For example, across cultures and history: Rhodes (2006), The evolutionary psychology of facial beauty.

198 Attractive faces are also symmetric: Rhodes (2006), The evolutionary psychology of facial beauty.

199 There's research showing that alcohol: Parker, L. C., Penton-Voak, I. S., Attwood, A. S., & Munafo, M. R., (2008). Effects of acute alcohol consumption on ratings of attractiveness of facial stimuli: Evidence of long-term encoding. *Alcohol and Alcoholism,* 43, 636–640.

199 But one less known effect of drinking is that: Souto, A., Bezerra, B. M., & Halsey, L. G. (2008). Alcohol intoxication reduces detection of asymmetry: An explanation for increased perceptions of facial attractiveness after alcohol consumption? *Perception,* 37, 955–958.

199 The research clearly shows that men like female: Rhodes (2006), The evolutionary psychology of facial beauty.

199 And in general, women prefer: Rhodes (2006), The evolutionary psychology of facial beauty.

199 When women are looking for short-term: Burriss, R. P., Rowland, H. M., & Little, A. C. (2008). Facial scarring enhances men's attractiveness for short-term relationships. *Personality and Individual Differences, 46*, 213–217.

199 It turns out that, in general, composite faces: Langlois, J. H., & Roggman, L. A. (1990). Attractive faces are only average. *Psychological Science, 1*, 115–121.

200 There is evidence, for example, that more attractive: Roberts, S. C., Little, A. C., Gosling, L. M., Perrett, D., & Carter, V. (2005). MHC-heterozygosity and human facial attractiveness. *Evolution and Human Behavior, 26*, 213–226.

200 Greater male facial attractiveness is also: Soler, C., Nunez, M., Gutierrez, R., Nunez, J., & Medina, P. (2003). Facial attractiveness in men provides clues to semen quality. *Evolution and Human Behavior, 24*, 199–207.

200 With regard to specific dimensions of attractiveness: Rhodes, G., Chan, J., Zebrowitz, L. A., & Simmons, L. W. (2003). Does sexual dimorphism in human faces signal health? *Proceedings of the Royal Society of London, Ser. B, 270*, S93–S95.

200 There's also a relation between facial averageness: Rhodes, G., Zebrowitz, L. A., Clark, A., Kalick, S., Hightower, A., & McKay, R. (2001). Do facial averageness and symmetry signal health? *Evolution and Human Behavior, 22*, 31–46.

200 Finally, it seems that in societies with less accessible: Rhodes (2006), The evolutionary psychology of facial beauty.

200 With regard to symmetry, it's well known that: Halberstadt, J., & Rhodes, G. (2000). The attractiveness of nonface averages: Implications for an evolutionary explanation of the attractiveness of average faces. *Psychological Sciences, 11*, 285–289.

201 With regard to our preference for average faces: Langlois, J. H., Roggman, L. A., & Musselman, L. (1994). What is average and what is not average about attractive faces? *Psychological Sciences, 5*, 214–20.

CHAPTER 9. *The Highest Form of Flattery*

204 In one experiment, fMRI imaging was: Scott, S. (2008). Imagination is the key to vocal mimicry. Presentation at the Society for Neuroscience, Washington, DC, November 15–19.

206 Research in the 1970s showed that students: LaFrance, M. (1979). Nonverbal synchrony and rapport: Analysis by the cross-lag panel technique. *Social Psychology Quarterly, 42*, 66–70.

207 A similar increase in posture matching: Charney, E. J. (1966). Postural configurations in psychotherapy, *Psychosomatic Medicine, 28*, 305–315.

207 In a seminal study, Tanya Chartrand and John Bargh: Chartrand, T. L., & Bargh, J. A. (1999). The chameleon effect: The perception-behavior link and social interaction, *Journal of Personality and Social Psychology, 76*, 893–910.

207 For example, not only do you adopt the types: Watanabe, K. (2008). Behavioral speed contagion: Automatic modulation of movement timing by observation of body movements. *Cognition, 106*, 1514–1524.

209 Research suggests that even while sitting: Paccalin, C., & Jeannerod, M. (2000). Changes in breathing during observation of effortful actions. *Brain Research, 862*, 194–200.

209 Research suggests that you inadvertently imitate the eating: Johnston, L. (2002). Behavioral mimicry and stigmatization. *Social Cognition, 20*, 18–35.

209 As you learned, by the time you were two weeks: Meltzoff, A. N., and Moore, M. K. (1977). Imitation of facial and manual gestures by human neonates. *Science, 198,* 75–78.

209 At that time, you also imitated facial postures: Chen, X., Striano, T., & Rakoczy, R. (2004). Auditory-oral matching behavior in newborns. *Developmental Science, 7,* 42–47.

209 Besides inadvertently imitating a person's accent: Pickering, M. J., & Garrod, S. (2004). Toward a mechanistic psychology of dialogue. *Behavioral and Brain Sciences, 27,* 169–22.

209 To some degree, you also mimic word choice: Pickering & Garrod (2004), Toward a mechanistic psychology of dialogue.

210 It turns out that when you produce a *p*: Shockley, K., Sabadini, L., & Fowler, C. A. (2004). Imitation in shadowing words. *Perception & Psychophysics, 66,* 422–429.

211 This was demonstrated in the lab in another: Chartrand & Bargh (1999), The chameleon effect.

211 Being subtly imitated by a "negotiator" would make you: Maddux, W. W., Mullen, E., & Glainsky, A. D. (2008). Chameleons bake bigger pies and take bigger pieces: Strategic behavioral mimicry facilitates negotiation outcomes. *Journal of Experimental Social Psychology, 44,* 461–468.

211 You'd give a higher tip to a waitress: van Baaren, R. B., Holland, R. W., Kawakami, K., & van Knippenberg, A. (2004). Mimicry and prosocial behavior. *Psychological Science, 15,* 71–74.

211 And you'd even rate a computer-animated "interviewer": Bailenson, J. N., & Yee, N. (2005). Digital chameleons: Automatic assimilation of nonverbal gestures in immersive virtual environments. *Psychological Science, 16,* 814–819.

211 Finally, when GPS driving directions are conveyed: Jonsson, I. M., Nass, C., Harris, H., & Takayama, L. (2008). Matching in-car voice with driver state: Impact on attitude and driving performance. *Proceedings of the Third International Driving Symposium on Human Factors in Driver Assessment, Training and Vehicle Design.* 173–180.

211 Imagine being in this experiment: van Baaren et al. (2004), Mimicry and prosocial behavior.

212 This would be true even if that motivation: Lakin, L., & Chartrand, T. L. (2003). Using nonconscious behavioral mimicry to create affiliation and rapport. *Psychological Science, 14,* 334–339.

212 If, for example, you felt left out of an online game: Lakin, J. L., Chartrand, T. L., & Arkin, R. M. (2008). I am too just like you: Nonconscious mimicry as an automatic behavioral response to social exclusion. *Psychological Science, 19,* 816–822.

212 There's also evidence that you do more imitating: Yabar, Y., Johnston, L., Miles, L., & Peace, V. (2006). Implicit behavioral mimicry: Investigating the impact of group membership. *Journal of Nonverbal Behavior, 30,* 97–113.

212 You sit in a chair, you're blindfolded, and then: Casile, A., & Giese, M. A. (2006). Nonvisual motor training influences biological motion perception, *Current Biology, 16,* 69–74.

213 If, for example, you were asked to determine: Reed, C. L., & Farah, M. J. (1995). The psychological reality of the body schema: A test with normal participants. *Journal of Experimental Psychology: Human Perception and Performance, 21,* 334–343.

214 Next, your ability to visibly judge the walking speed: Jacobs, A., and Shiffrar, M. (2005). Walking perception by walking observers. *Journal of Experimental Psychology: Human Perception & Performance, 31,* 157–169.

214 And your visual judgments of the weight of a small box: Hamilton, A., Wolpert, D., & Frith, U. (2004). Your own action influences how you perceive another person's action. *Current Biology, 14,* 493–498.

214 Imagine being in this menacing experiment: Ito, T., Tiede, M., & Ostry, D. J. (2009). Somatosensory function in speech perception. *Proceedings of the National Academy of Sciences, USA, 106*, 1245–1248.

214 If, for example, you silently articulate: Sams, M., Mottonen, R., & Sihvonen, T. (2005). Seeing and hearing others and oneself talk. *Cognitive Brain Research, 23*, 429–435.

215 If, for example, you're looking at an experimenter: Fadiga, L., Fogassi, L., Pavesi, G., & Rizzolatti, G. (1995). Motor facilitation during action observation: A magnetic stimulation study. *Journal of Neurophysiology, 73*, 2608–2611.

215 If, instead, you were watching the experimenter: Fadiga et al. (1995), Motor facilitation during action observation.

215 By recording from the muscles: Fadiga et al. (1995), Motor facilitation during action observation.

216 Listening to manual actions such as paper: Aziz-Zadeh, L., Iacoboni, M., Zaidel, E., Wilson, S., & Mazziotta, J. (2004). Left hemisphere motor facilitation in response to manual action sounds. *European Journal of Neuroscience, 19*, 2609–2612.

216 Hearing syllables containing tongue-tip: Fadiga, L., Craighero, L., Buccino, G., & Rizzolatti, G. (2002). Speech listening specifically modulates the excitability of tongue muscles: A TMS study. *European Journal of Neuroscience, 15*, 399–402.

217 Giacomo Rizzolatti and his colleagues: Rizzolatti, G., Fadiga, L., Gallese, V., & Fogassi, L. (1995). Premotor cortex and the recognition of motor actions. *Cognitive Brain Research, 3*, 131–141.

217 We now know that there are different subtypes: Rizzolatti, G., & Craighero, L. (2004). The mirror-neuron system. *Annual Review of Neuroscience, 27*, 169–192.

217 Also, most mirror neurons will only: Iacoboni, M., & Dapretto, M. (2006). The mirror neuron system and the consequences of its dysfunction. *Nature Reviews Neuroscience, 7*, 942–951.

218 The imaging shows that regions you use: Iacoboni & Dapretto (2006), The mirror neuron system.

218 There is also evidence that *disrupting* specific: Iacoboni & Dapretto (2006), The mirror neuron system.

218 If, for example, a transient lesion were produced: Pobric, G., & Hamilton, A. F. (2006). Action understanding requires the left inferior frontal cortex. *Current Biology, 16*, 524–529.

218 Relatedly, a transient lesion induced in another: Urgesi, C., Candidi, M., Ionta, S., & Aglioti, S. M. (2007). Representation of body identity and body actions in extrastriate body area and ventral premotor cortex. *Nature Reviews Neuroscience, 10*, 30–31.

218 If a transient lesion were induced in another: Heiser, M., Iacoboni, M., Maeda, F., Marcus, J., & Mazziotta, J. C. (2003). The essential role of Broca's area in imitation. *European Journal of Neuroscience, 17*, 1123–1128.

218 In fact, brain imaging shows that while these areas: Iacoboni & Dapretto (2006), The mirror neuron system.

219 For example, it's been thought that dysfunctions in the mirror system: Iacoboni & Dapretto (2006), The mirror neuron system.

219 In fact, recent brain scans have shown that: Hadjikhani, N., Joseph, R. M., Snyder, J., & Tager-Flusberg, H. (2006). Anatomical differences in the mirror neuron system and social cognition network in autism. *Cerebral Cortex, 16*, 1276–1282.

219 Other brain scan research revealed that autistic: Nishitani, N., Avikainen, S., & Hari, R. (2004). Abnormal imitation-related cortical activation sequences in Asperger's syndrome. *Annals of Neurology, 55*, 558–562.

220 Still, some provocative new brain imaging evidence: Gazzola, V., Aziz-Zadeh, L.,

& Keysers, C. (2006). Empathy and the somatotopic auditory mirror system in humans. *Current Biology, 16,* 1824–1829.

220 Finally, there is speculation that the mirror: Arbib, M. A. (2005). From monkey-like action recognition to human language: An evolutionary framework for neurolinguistics. *Behavioral and Brain Sciences, 28,* 105–167.

223 You walk into a laboratory room: Tatler, B. W., & Kuhn, G. (2007). Don't look now: The magic of misdirection. In R.P.G. van Gompel, M. H. Fischer, W. S. Murray, & R. L. Hill (Eds). *Eye movements: A window on mind and brain* (697–714). Amsterdam: Elsevier Science.

224 You sit in front of a computer monitor, and again: Ricciardelli, P., Bricolo, E., Aglioti, S. M., & Chelazzi, L. (2002). My eyes want to look where your eyes are looking: Exploring the tendency to imitate another individual's gaze. *NeuroReport, 13,* 2259–2264.

225 If a similar experiment were conducted: Sato, W., Okada, T., and Toichi, M. (2007). Attentional shift by gaze is triggered without awareness. *Experimental Brain Research, 183,* 87–94.

225 You look at a computer screen on which appears: Bayliss, A. P., Paul, M. A., Cannon, P. R., & Tipper, S. P. (2006). Gaze cueing and affective judgments of objects: I like what you look at. *Psychonomic Bulletin & Review, 13,* 1061–1066.

226 It turns out that without knowing why: Bayliss, A. P., & Tipper, S. P. (2006). Predictive gaze cues and personality judgments: Should eye trust you? *Psychological Science, 17,* 514–520.

226 You put on an eye-tracking cap: Kuhn, G., & Land, M. F. (2006). There's more to magic than meets the eye. *Current Biology, 16,* R950–R951.

229 You sit at a table on which is propped: Oullier, O., de Guzman, G. C., Jantzen, K. J., Lagarde, J., & Kelso, J.A.S. (2008). Social coordination dynamics: Measuring human bonding. *Social Neuroscience, 3,* 178–192.

230 For example, if you were asked to sit in a weighted: Richardson, M. J., Marsh, K. J., Isenhower, R. W., Goodman, J.R.L., & Schmidt, R. C. (2007). Rocking together: Dynamics of intentional and unintentional interpersonal coordination. *Human Movement Science, 26,* 867–891.

230 This has long been known about crickets: Walker, T. J. (1969). Acoustic synchrony: Two mechanisms in the snow tree cricket. *Science, 166,* 891–894.

230 A human example of auditory-based synchronization: Neda, Z., Ravasz, E., Brechet, Y., Vicsek, T., & Barabasi, A. L. (2000). The sound of many hands clapping: Tumultuous applause can transform itself into waves of synchronized clapping. *Nature, 403,* 849–850.

230 You're asked to perform a "find the differences": Shockley, K., Santana, M. V., & Fowler, C. A. (2003). Mutual interpersonal postural constraints are involved in cooperative conversation. *Journal of Experimental Psychology: Human Perception and Performance, 29,* 326–332.

231 In general, it is the case that the opposite-phase: Kelso, J.A.S. (1984). Phase transitions and critical behavior in human bimanual coordination. *American Journal of Physiology: Regulatory, Integrative, and Comparative, 246,* R1000–R1004; Turvey, M. T., Rosenblum, L. D., Schmidt, R. C., & Kugler, P. N. (1986). Fluctuations and phase symmetry in coordinated rhythmic movements. *Journal of Experimental Psychology: Human Perception & Performance, 12,* 564–583.

232 When your limbs coordinate, there's an important: Kugler, P. N., & Turvey, M. T. (1987). *Information, natural law, and the self-assembly of rhythmic movement.* Hillsdale, NJ: Lawrence Erlbaum Associates, Inc.

232 Evidence for this has come from experiments: Schmidt, R. C., Carello, C., & Turvey, M. T. (1990). Phase transitions and critical fluctuations in the visual coordination of rhythmic movements between people. *Journal of Experimental Psychology: Human Perception and Performance, 16,* 227–247.

233 And this difference also determines: Schmidt, R. C., & Turvey, M. T. (1994). Phase-entrainment dynamics of visually coupled rhythmic movements. *Biological Cybernetics, 70,* 369–376.

233 For example, there is evidence that even without touching: Bernieri, F. J., Reznick, S., & Rosenthal, R. (1988). Synchrony, pseudosynchrony, and dissynchrony: Measuring the entrainment process in mother-infant interactions, *Journal of Personality and Social Psychology, 54,* 243–253.

233 Also, you're much more likely to inadvertently: de Guzman, G. C., Tognoli, E., Lagarde, J., Jantzen, K. J., & Kelso, J.A.S. (2005). Effects of biological relevance of the stimulus in mediating spontaneous visual social coordination. *Society for Neuroscience: Abstract Viewer/Itinerary Planner,* Program No. 867.21.

233 You sit, and are asked to put your right forearm: Macrae, C. N., Duffy, O. K., Miles, L. K., & Lawrence, J. (2008). A case of hand waving: Action synchrony and person perception. *Cognition, 109,* 152–156.

235 Imagine being told that your task: Wiltermuth, S. S., & Heath, C. (2009). Synchrony and Cooperation, *Psychological Science, 20,* 1–5.

CHAPTER 10. *See What I'm Saying*

242 While lipreading skill varies widely: Auer, E., & Bernstein, L. (2007). Enhanced visual speech perception in individuals with early-onset hearing impairment. *Journal of Speech, Language, and Hearing Research, 50,* 1157–1165.

242 Research shows a link between superior: Auer & Bernstein (2007), Enhanced visual speech perception in individuals.

242 The research also suggests that being a woman: Auer & Bernstein (2007), Enhanced visual speech perception in individuals.

242 Finally, Thomas is a very bright person: Auer & Bernstein (2007), Enhanced visual speech perception in individuals.

243 For these individuals, the visual brain: Bavelier, D., Dye, M. W., & Hauser, P. C. (2006). Do deaf individuals see better? *Trends in Cognitive Sciences, 10,* 512–518.

243 Early-deaf individuals' visual brain areas: Bavelier et al. (2006), Do deaf individuals see better?

243 These regions have been shown to respond: Sadato, N., Okada, T., Honda, M., Matsuki, K., Yoshida, M., & Kashikura, K. (2005). Cross-modal integration and plastic changes revealed by lip movement, random-dot motion and sign languages in the hearing and deaf. *Cerebral Cortex, 15,* 1113–1122.

243 The early-deaf are faster and more accurate: Bavelier et al. (2006), Do deaf individuals see better?

243 This is especially true with motion: Bavelier, D., Brozinsky, C., Tomann, A., Mitchell, T., Neville, H., & Liu, G. (2001). Impact of early deafness and early exposure to sign language on the cerebral organization for motion processing. *Journal of Neuroscience, 21,* 8931–8942.

244 Most recent estimates show that: Agrawal, Y., Platz, E. A., & Niparko, J. K. (2008). Prevalence of hearing loss and differences by demographic characteristics among US adults: Data from the national health and nutrition examination survey, 1999–2004. *Archives of Internal Medicine, 168,* 1522–1530.

245 This phenomenon is known as the McGurk effect: McGurk, H., & MacDonald, J. W. (1976). Hearing lips and seeing voices. *Nature, 264,* 746–748.

245 In our own laboratory, we find that: Rosenblum, L. D., and Saldaña, H. M. (1996). An audiovisual test of kinematic primitives for visual speech perception. *Journal of Experimental Psychology: Human Perception and Performance. 22,* 318–331.

246 It shouldn't matter: knowing that the voice you hear: Green, K. P., Kuhl, P. K., Meltzoff, A. M., & Stevens, E. B. (1991). Integrating speech information across talkers, gender, and sensory modality: Female faces and male voices in the McGurk effect. *Perception & Psychophysics, 50,* 524–536.

246 The McGurk effect works with speech perceivers: Sekiyama, K., & Tokhura, Y. (1993). Inter-language differences in the influence of visual cues in speech perception. *Journal of Phonetics, 21,* 427–444.

246 The effect works whether you're looking: Jordan, T. R., & Sergeant, P. [C.] (2000). Effects of distance on visual and audiovisual speech recognition. *Language & Speech, 43,* 107–124; Jordan, T. R., & Thomas, S. M. (2001). Effects of horizontal viewing angle on visual and audiovisual speech recognition. *Journal of Experimental Psychology: Human Perception & Performance, 27,* 1386–1403.

246 Interestingly, the effect is sometimes reduced: Jordan, T. R., & Bevan, K. (1997). Seeing and hearing rotated faces: Influences of facial orientation on visual and audiovisual speech recognition. *Journal of Experimental Psychology: Human Perception and Performance, 23,* 388–403.

246 Our own lab has shown that the effect works: Rosenblum & Saldaña (1996), An audiovisual test of kinematic primitives.

247 Research in our lab and others shows: Rosenblum, L. D., Johnson, J. A., & Saldaña, H. M. (1996). Visual kinematic information for embellishing speech in noise. *Journal of Speech and Hearing Research, 39,* 1159–1170.

247 Careful analysis of facial motions: Munhall, K., & Vatikiotis-Bateson, E. (2004). Spatial and temporal constraint on audiovisual speech perception. In G. A. Calvert, C. Spence, & B. E. Stein (Eds.) *The handbook of multisensory processes* (pp. 177–188). Cambridge, MA: MIT Press.

248 It turns out that the very quick air pressure: Munhall, K., & Vatikiotis-Bateson E. (2004), Spatial and temporal constraint on audiovisual speech perception.

248 In one experiment, a talker was videotaped: Munhall, K. G., Jones, J. A., Callan, D. E., Kuratate, T., & Vatikiotis-Bateson, E. (2004). Visual prosody and speech intelligibility head movement improves auditory speech perception. *Psychological Science, 15,* 133–137.

249 Other research shows that you can distinguish: Burnham, D., Ciocca, V., Lauw, C., Lau, S., & Stokes, S. (2000). Perception of visual information for Cantonese tones. In M. Barlow & P. Rose (Eds). *Proceedings of the Eighth Australian International Conference on Speech Science and Technology* (pp. 86–91). Canberra: Australian Speech Science and Technology Association.

249 But even when asked to explicitly perceive: Vatikiotis-Bateson, E., Eigsti, I.-M., Yano, S., & Munhall, K. G. (1998). Eye movement of perceivers during audiovisual speech perception. *Perception & Psychophysics, 60,* 926–940.

250 You can't induce a real McGurk effect: Fowler & Dekle (1991), Listening with eye and hand.

251 We were interested in whether five-month-olds: Rosenblum, L. D., Schmuckler, M. A., & Johnson, J. A. (1997). The McGurk effect in infants. *Perception & Psychophysics, 59,* 347–357.

253 Since our experiment, other labs have replicated: Burnham, D., & Dodd, B. (2004).

Auditory-visual speech integration by prelinguistic infants: Perception of an emergent consonant in the McGurk effect. *Developmental Psychobiology, 45,* 204–220.

253 Another study used brain imaging: Kushnerenko, E., Teinonen, T., Volein, A., & Csibra, G. (2008). Electrophysiological evidence of illusory audiovisual speech percept in human infants. *Proceedings of the National Academy of Sciences USA, 105,* 442–445.

253 But research using other methods indicates: Rosenblum et al. (1997). The McGurk effect in infants.

253 One recent experiment demonstrated that infants: Teinonen, T., Aslin, R. N., Alku, P., & Csibra, G. (2008). Visual speech contributes to phonetic learning in 6-month-old infants. *Cognition, 108,* 850–855.

253 Many blind infants acquire language: Mills, A. E. (1987). The development of phonology in the blind child. In B. Dodd & R. Campbell (Eds.), *Hearing by eye: The psychology of lip reading* (pp. 145–162). Hillsdale, NJ: Lawrence Erlbaum Associates, Inc.

254 By the time you were four months: Weikum, W., Vouloumanos, A., Navarro, J., Soto-Faraco, S., Sebastian-Galles, N., & Werker, J. F. (2007). Visual language discrimination in infancy. *Science, 316,* 1159.

254 Interestingly, however, there was one time: Massaro, D. W., Thompson, L. A., Barron, B., & Laren, E. (1986). Developmental changes in visual and auditory contributions to speech perception. *Journal of Experimental Child Psychology, 41,* 93–113.

254 It turns out that when you are presented: Mottonen, R., Krause, C. M., Tiippana, K., Sams, M. (2002). Processing of changes in visual speech in the human auditory cortex. *Cognitive Brain Research, 13,* 417–425.

255 In the original demonstration of this phenomenon: Calvert, G. A., Bullmore, E., Brammer, M. J., Campbell, R., Iversen, S. D., Woodruff, P., McGuire, P., Williams, S., David, A. S. (1997). Silent lipreading activates the auditory cortex. *Science, 276,* 593–596.

255 For example, there is now evidence that the visual: Kislyuk, D. S., Mottonen, R., & Sams, M. (2008). Visual processing affects the neural basis of auditory discrimination. *Journal of Cognitive Neuroscience, 20,* 2175–2184.

255 Other research has shown that lip-read speech: Musacchia, G., Sams, M., Nicol, T., & Kraus, N. (2005). Seeing speech affects acoustic information processing in the human brainstem. *Experimental Brain Research, 168,* 1–10.

255 Specifically, if you were presented an ambiguous: Green, K. P., & Miller, J. L. (1985). On the role of visual rate information in phonetic perception. *Perception and Psychophysics, 38,* 269–276.

256 In the context of the McGurk effect: Ojanen, V., Mottonen, R., Pekkola, J., Jaaskelainen, I. P., Joensuu, R., Autti, T., & Sams, M. (2005). Processing of audiovisual speech in Broca's area. *NeuroImage 25,* 333–338

256 Interestingly, activation of motor areas: Santi A., Servos, P., Vatikiotis-Bateson, E., Kuratate, T., Munhall, K. (2003). Perceiving biological motion: Dissociating visible speech from walking. *Journal of Cognitive Neuroscience, 15,* 800–809.

256 If you were to have your facial motor: Sundara, M., Namasivayam, A. K., Chen, R. (2001). Observation-execution matching system for speech: A magnetic stimulation study. *NeuroReport, 12,* 1341–1344.

256 This research used the TMS pulse technique: Watkins, K. E., Strafella, A. P., & Paus, T. (2003). Seeing and hearing speech excites the motor system involved in speech production. *Neuropsychologia, 41,* 989–994.

257 Imagine being in this experiment, conducted in our lab: Miller, R. M., Rosenblum, L. D., & Sanchez, K. (2006). Phonetic alignment to visual speech. *Journal of the Acoustical Society of America, 120*(5), 3249.

257 In another experiment conducted in our lab: Sanchez, K., Miller, R. M., &

Rosenblum, L. D. (in press). Visual influences on alignment to voice onset time. *Journal of Speech, Language, and Hearing Research*.

260 If, for example, you listened to a single word: Kamachi, M., Hill, H., Lander, K., & Vatikiotia-Bateson, E. (2003). Putting the face to the voice: Matching identity across modality. *Current Biology, 13,* 1709–1714.

261 You could make the voice-face matches for talkers: Lander, K., Hill, H., Kamachi, M., Vatikiotis-Bateson, E. (2007). It's not what you say but the way you say it: Matching faces and voices. *Journal of Experimental Psychology: Human Perception and Performance, 33,* 905–914.

261 Work in our lab has shown that you: Rosenblum, L. D., Smith, N. M., Nichols, S., Lee, J., & Hale, S. (2006). Hearing a face: Cross-modal speaker matching using isolated visible speech. *Perception & Psychophysics, 68,* 84–93.

262 In experiments conducted in another laboratory: Lachs, L., & Pisoni, D. B. (2004). Specification of cross-modal source information in isolated kinematic displays of speech. *Journal of the Acoustical Society of America,* 116, 507–518.

263 This is completely consistent with research conducted: Rosenblum, L. D. (2005). The primacy of multimodal speech perception. In D. Pisoni & R. Remez (Eds.), *Handbook of speech perception* (pp. 51–78). Malden, MA: Blackwell.

263 Research shows that as you become more: Nygaard, L. C. (2005). The integration of linguistic and non-linguistic properties of speech. In D. Pisoni & R. Remez (Eds.), *Handbook of speech perception* (pp. 390–414). Malden, MA: Blackwell.

263 This led our lab to make one of the wildest predictions: Rosenblum, L. D., Miller, R. M., Sanchez, K. (2007). Lip-read me now, hear me better later: Cross-modal transfer of talker-familiarity effects. *Psychological Science, 18,* 392–396.

264 This is supported by evidence that monkeys: Ghanzafar, A. A., & Logothetis, N. K. (2003). Facial expressions linked to monkey calls. *Nature, 423,* 937–938.

264 And for integrating audiovisual calls: Ghazanfar, A. A., Maier, J. X., Hoffman, K. L., & Logothetis, N. K. (2005). Multisensory integration of dynamic faces and voices in rhesus monkey auditory cortex. *Journal of Neuroscience, 25,* 5004–5012.

264 There is also speculation that the world's languages: Rosenblum (2005), The primacy of multimodal speech perception.

CHAPTER 11. *All of the Above*

268 In the aptly named *ventriloquist effect*: Vroomen, J., & de Gelder, B. (2004). Perceptual effects of cross-modal stimulation: Ventriloquism and the freezing phenomenon. In G. Calvert, C. Spence, & B. E. Stein (Eds.), *The handbook of multisensory processes* (pp. 141–150). Cambridge, MA: MIT Press.

268 Other research shows that the number of simple flashes: Shams, L., Kiamitani, Y., & Shimojo, S. (2000). What you see is what you hear. *Nature, 408,* 788.

268 In another blip-based effect: Sekuler, R., Sekuler, A. B., & Lau, R. (1997). Sound alters visual motion perception. *Nature, 385,* 308.

268 For example, being touched just as you see: Watanabe, R., & Shimojo, S. (1998). Attentional modulation in perception of visual motion events. *Perception, 27,* 1041–1054.

268 In a study reminiscent of the potato chip: Jousmäki, V., & Hari, R. (1998). Parchment-skin illusion: Sound-biased touch. *Current Biology, 8,* R190.

269 Your sense of a sound's horizontal location: Caclin, A., Soto-Faraco, S., Kingstone, A., & Spence, C. (2002). Tactile "capture" of audition. *Perception & Psychophysics, 64,* 616–630.

269 In our lab, we found that the sight: Saldaña, H. M., and Rosenblum, L. D. (1993). Visual influences on auditory pluck and bow judgments. *Perception and Psychophysics, 54,* 406–416.

269 Next, the sight of someone exaggerating: Schutz, M., & Lipscomb, S. (2007). Hearing gestures, seeing music: Vision influences perceived tone duration. *Perception, 36,* 888–897.

269 Also, the pitch difference heard between: W. F., Graham, P., & Russo, F. A. (2005). Seeing music performance: Visual influences on perception and experience. *Semiotica, 156,* 203–227.

269 And in another study conducted in our lab: Rosenblum, L. D., and Fowler, C. A. (1991). Audio-visual investigation of the loudness-effort effect for speech and nonspeech stimuli. *Journal of Experimental Psychology: Human Perception and Performance, 17,* 976–985.

269 If you look steadily at a visual grating: Konkle, T., Wang, Q., Hayward, V., & Moore, C. I. (2009). Motion aftereffects transfer between touch and vision. *Current Biology, 19,* 1–6.

270 Having your calf muscle subtly stimulated: Lugo, J. E., Doti, R., Wittich, W., Faubert, J. (2008). Multisensory integration central processing modifies peripheral systems. *Psychological Science, 19,* 989–997.

272 You're told to walk into what looks like: Lee, D. N., & Lishman, J. R. (1977). Vision—the most efficient source of proprioceptive information for balance control. *Agressologie, 18,* 83–94.

274 It turns out that a relativity small visual image: Andersen, G. J., & Dyre, B. P. (1989). Spatial orientation from optic flow in the central visual field. *Perception & Psychophysics, 45,* 453–458.

274 Even toddlers are influenced by visual flow: Lee, D. N., & Aronson, E. (1974). Visual proprioceptive control of standing in human infants. *Perception & Psychophysics, 15,* 529–532.

274 In an experiment that did not involve: Stoffregen, T. A., Smart, L. J., Bardy, B. G., & Pagulayan, R. J. (1999). Postural stabilization of looking. *Journal of Experimental Psychology: Human Perception and Performance, 25,* 1641–1658.

274 If you were blindfolded and heard sounds: Lackner, J. R., (1997). Induction of illusory self-rotation and nystagmus by a rotating sound-field. *Aviation, Space, and Environmental Medicine, 48,* 129–131; Sakamoto, S., Osada, Y., Suzuki, Y., & Gyoba, J. (2004). The effects of linearly moving sound images on self motion perception. *Acoustical Science and Technology, 25,* 100–102.

275 If, for example, you were hearing sounds shift: Valjamae, A., Larsson, P., Vastfjall, D., & Kleiner, M. (2006). Vibrotactile enhancement of auditory-induced self-motion and spatial presence. *Audio Engineering Society, 54,* 954–963.

275 A group of strippers anonymously provided: Miller, G. F., Tybur, J., & Jordan, B. (2007). Ovulatory cycle effects on tip earnings by lap-dancers: Economic evidence for human estrus? *Evolution and Human Behavior, 28,* 375–381.

275 The other studies showed that men act more jealous: Haselton, M. G., & Gangestad, S. W. (2006). Conditional expression of women's desires and men's mate guarding across the ovulatory cycle. *Hormones and Behavior, 49,* 509–518.

276 It turns out that during her estrus: Roberts, S. C., Havlicek, J., Flegr, J., Hruskova, M., Little, A. C. (2004) Female facial attractiveness increases during the fertile phase of the menstrual cycle. *Proceedings of the Royal Society of London, Ser. B, 271,* S270–S272.

276 Women during estrus also show a subtly: Kirchengast, S., & Gartner, M. (2002). Changes in fat distribution (WHR) and body weight across the menstrual cycle. *Collegium Anthropologicum, 26,* S47–S57.

276 Thirty college-aged women were photographed: Haselton, M. G., Mortezaie, M., Pillsworth, E. G., Bleske, A. E., & Frederick, D. A. (2007). Ovulatory shifts in human female ornamentation: Near ovulation, women dress to impress. *Hormones and Behavior, 51,* 40–45.

277 In these studies, women are recorded during: Pipitone, R. N., & Gallup, G. G. (2008). Women's voice attractiveness varies across the menstrual cycle. *Evolution and Human Behavior, 29,* 268–274.

277 Further, there is evidence that women in estrus: Symonds, C. S., Gallagher, P., Thompson, J. M., & Young, A. H. (2004). Effects of the menstrual cycle on mood, neurocognitive and neuroendocrine function in healthy premenopausal women. *Psychological Medicine, 34,* 93–102.

277 They also self-report more flirtatious behavior: Gangestad, S. W., Thornhill, R., & Garver, C. E., (2002). Changes in women's sexual interests and their partner's mate retention tactics across the menstrual cycle: Evidence for shifting conflicts of interest. *Proceedings of the Royal Society of London, Ser. B, 269,* 975–982.

278 But these visible dimensions have recently: Collins, S. A., & Missing, C. (2003). Vocal and visual attractiveness are related in women. *Animal Behaviour, 65,* 997–1004.

278 These same utterances also related: Hughes, S. M., Dispenza, F., & Gallup, G. G. (2004). Ratings of voice attractiveness predict sexual behavior and body configuration. *Evolution and Human Behavior, 25,* 295–304.

278 And other research has found a relationship: Rikowski, A., & Grammer, K. (1999). Human body odour, symmetry and attractiveness. *Proceedings of the Royal Society of London, Ser. B, 266,* 869–874.

278 To determine how voice attractiveness: Feinberg, D. R., Jones, B. C., DeBruine, L. M., Moore, F. R., Law Smith, M. J. (2005). The voice and face of woman: One ornament that signals quality? *Evolution and Human Behavior, 26,* 398–408.

278 There's evidence that the common characteristics: Feinberg et al. (2005), The voice and face of a woman.

279 For example, it turns out that if you are: Despres, O., Candas, T., & Dufour, A. (2005). Auditory compensation in myopic humans: Involvement of binaural, monaural, or echo cues? *Brain Research, 1041,* 56–65.

279 In a particularly stunning example of cross-sensory brain: Von Melchner, L., Pallas, S. L., & Sur, M. (2000). Visual behaviour mediated by retinal projections directed to the auditory pathway. *Nature, 404,* 871–876.

280 For example, the presence of visible events: Stekelenburg, J. J., & Vroomen, J. (2007). Neural correlates of multisensory integration of ecologically valid audiovisual events. *Journal of Cognitive Neuroscience, 19,* 1964–1973.

280 In fact, transient brain lesions: Macaluso, E. (2006). Multisensory processing in sensory-specific cortical areas. *The Neuroscientist, 12,* 327–338.

280 Related research has shown that *identifying:* Macaluso (2006), Multisensory processing in sensory-specific cortical areas.

280 Other experiments show that the *position:* Macaluso (2006), Multisensory processing in sensory-specific cortical areas.

281 It could be that in an important way: Rosenblum, L. D. (2008). Speech perception as a multimodal phenomenon. *Current Directions in Psychological Science, 17,* 405–409.

281 as the ball gets closer, the *rate of change:* DeLucia, P. R. (2004). Time-to-contact judgments of an approaching object that is partially concealed by an occluder. *Journal of Experimental Psychology: Human Perception and Performance, 30,* 287–304.

281 In this way, the relevant information is the same: Lee, D., van der Weel, F. R., Hitchcock, T., Matejowssky, E., & Pettigrew, J. (1992). Common principle of guidance by echolocation and vision. *Journal of Comparative Physiology, Ser. A, 171,* 563–571.

281 We conducted an initial experiment: Gordon, M. S., & Rosenblum, L. D. (2005). Effects of Intra-Stimulus Modality Change on Audiovisual Time-to-Arrival Judgments. *Perception & Psychophysics, 67,* 580–594.

282 The idea that the relevant perceptual information: Rosenblum (2005), The primacy of multimodal speech perception; Summerfield, Q. (1987). Some preliminaries to a comprehensive account of audio-visual speech perception. In B. Dodd & R. Campbell (Eds.), *Hearing by eye: The psychology of lip-reading* (pp. 53–83). Hillsdale, NJ: Lawrence Erlbaum Associates, Inc.

282 Measurements taken from the front: Yehia, H. C., Rubin, P. E., and Vatikiotis-Bateson, E. (1998) Quantitative association of vocal-tract and facial behavior. *Speech Communication, 26,* 23–43.

282 Other similar examples of cross-sensory: Rosenblum (2005), The primacy of multimodal speech perception.

283 For many synesthetes: Ramachandran, V. S., & Hubbard, E. M. (2001). Synaesthesia: A window into perception, thought, and language. *Journal of Consciousness Studies, 8*(12), 3–34.

284 The estimated number of individuals: Simner, J., Mulvenna, C., Sagiv, N., Tsakanikos, E., Witherby, S. A., Fraser, C., Scott, K., & Ward, J. (2006). Synaesthesia: The prevalence of atypical cross-modal experiences. *Perception, 35,* 1024–1033.

284 In fact, synesthetes in the general population: Ramachandran & Hubbard (2001), Synaesthesia: A window into perception, thought, and language.

284 These include the artists: http://en.wikipedia.org/wiki/Synesthesia, Retrieved May, 15, 2009.

284 Early explanations of synesthesia: Ramachandran & Hubbard (2001), Synaesthesia: A window into perception, thought, and language.

284 Synesthetes perform differently on perceptual: Ramachandran, V. S., Hubbard, E. M. (2001). Psychophysical investigations into the neural basis of synaesthesia. *Proceedings of the Royal Society of London, Ser. B, 268,* pp. 979–83.

284 When presented letters or numbers: Aleman, A., Rutten, G. M., Sitskoorn, M. M., Dautzenberg, G., Ramesy, N. F. (2001). Activation of striate cortex in the absence of visual stimulation: An fMRI study of synesthesia. *NeuroReport, 12,* 2827–2830.

284 One current theory is that synesthetes: Bargary, G., & Mitchell, K. J. (2008). Synaesthesia and cortical connectivity. *Trends in Neurosciences, 31,* 335–342.

284 Another theory is that synesthetes: Cohen Kadosh, R., & Walsh, V. (2008). Synaesthesia and cortical connections: Cause or correlation? *Trends in Neurosciences, 31,* 549–550.

285 Very recent examinations: Witthoft, N., & Winawer, J. (2006). Synesthetic colors determined by having colored refrigerator magnets in childhood. *Cortex, 42,* 175–183.

285 In fact, it turns out that *many* letter-color synesthetes: Simner, J., Ward, J., Lanz, M., Jansari, A., Noonan, K., Glover, L., & Oakley, D. (2005). Non-random associations of graphemes to colours in synaesthetic and normal populations. *Cognitive Neuropsychology, 22,* 1069–1085.

285 For example, it turns out that many of the: Simner et al. (2005), Non-random associations of graphemes to colours.

286 If you were asked to choose colors: Ward, J., Huckstep, B., & Tsakanikos, E. (2006). Sound-colour synaesthesia: To what extent does it use cross-modal mechanisms common to us all? *Cortex, 42,* 264–280.

286 Of course, your specific selections: Ward et al. (2006), Sound-colour synaesthesia.

286 In fact, the artistic advantage of synesthetes: Ward, J., Moore, S., Thompson-Lake, D., Salih, S., & Beck, B. (2008). The aesthetic appeal of auditory-visual synaesthetic perceptions in people without synaesthesia. *Perception, 37,* 1285–1296.

287 Imagine being in this spooky experiment: Cohen Kadosh, R., Hnik, A., Catena, A., Walsh, V., & Fuentes, L. J. (2009). Induced cross-modal synaesthetic experience without abnormal neuronal connections. *Psychological Science, 20,* 258–265.

288 By some accounts, this is a cross-sensory: Steveson, R. J., & Tomiczek, C. (2007). Olfactory-induced synesthesias: A review and model. *Psychological Bulletin, 133,* 294–309.

288 The synesthetic tendencies in all of us: Martino, G., & Marks, L. E. (2000). Synesthesia: Strong and weak. *Current Directions in Psychological Science, 10,* 61–65.

289 For example, in most languages: Maurer, D., & Mondloch, C. J. (2005). Neonatal synesthesia: A reevaluation. In L. C. Robertson & N. Sagiv (Eds.), *Synesthesia: Perspectives from cognitive neuroscience* (pp. 193–213). Oxford: Oxford University Press.

290 If you're like 95 percent of people asked: Ramachandran & Hubbard (2001), Synaesthesia: A window into perception, thought, and language.

290 In fact, you would have made these same: Maurer, D., Pathman, T., & Mondlock, C.J. (2006). The shape of boubas: Sound–shape correspondences in toddlers and adults. *Developmental Science, 9,* 316–322.

290 It could very well be that to your: Ramachandran & Hubbard (2001), Synaesthesia: A window into perception, thought, and language.

290 Similar correspondences have been found: Maurer et al. (2006), The shape of boubas.

291 According to Ramachandran: Ramachandran & Hubbard (2001), Synaesthesia: A window into perception, thought, and language.

291 In fact, the supposed arbitrary relation: Gasser, M. (2004). The origins of arbitrariness in language. In *Proceedings of the Cognitive Science Society* (pp. 434–439). Hillsdale, NJ: Lawrence Erlbaum Associates, Inc.

291 In many languages, words referring to little things: Ramachandran & Hubbard (2001), Synaesthesia: A window into perception, thought, and language.

291 Going a step further, Ramachandran: Ramachandran & Hubbard (2001), Synaesthesia: A window into perception, thought, and language.

292 In one study, non-Japanese speaking subjects: Kunihira, S. (1971). Effects of the expressive voice on phonetic symbolism. *Journal of Verbal Learning and Verbal Behavior, 10,* 427–429.

292 And very recent research suggests that your: Nygaard, L. C., Cook, A. E., & Namy, L. L. (2009). Sound to meaning correspondences facilitate word learning. *Cognition 112,* 181–186.

Photograph Credits

Chapter 10, Point-Light Speech Technique
Photo: Lawrence Rosenblum; Model: Rachel Miller

Chapter 11, Moving Room Technique
Photo: Ken Jones; Model: Madeleine Dzinas Schmuckler

Index

Page numbers in *italics* refer to illustrations.